JOE NAGATA'S
LEGO® MINDSTORMS™ IDEA BOOK

JOE NAGATA

with illustrations by John VanZwieten

NO STARCH PRESS

San Francisco

JOE NAGATA'S
LEGO MINDSTORMS
IDEA BOOK

Printed in Canada

1 2 3 4 5 6 7 8 9 10—04 03 02 01

Publisher: William Pollock
Project Editor: Karol Jurado
Assistant Editor: Nick Hoff
Editorial Production Assistant: Jennifer Arter
Cover and Interior Design: Octopod Studios
Composition: Magnolia Studio and Octopod Studios
Copyeditor: Carol Lombardi
Proofreader: Ruth Stevens

Joe Nagata's LEGO MINDSTORMS Idea Book is an English version of *Joe Nagata no LEGO MindStorms Supa Kuricha,* the Japanese original edition, published in Japan by Ohmsha (Tokyo), copyright © 1999 by Joe Nagata. English translation prepared by Arnie Rusoff.

Distributed to the book trade in the United States by Publishers Group West, 1700 Fourth Street, Berkeley, California 94710, phone: 800-788-3123 or 510-528-1444, fax: 510-528-3444.

Distributed to the book trade in Canada by Jacqueline Gross & Associates, Inc., 195 Allstate Parkway, Markham, Ontario L3R 4T8 Canada, phone: 905-477-0722, fax: 905-477-8619.

For information on translations or book distributors outside the United States and Canada, please contact No Starch Press directly:

No Starch Press
555 De Haro Street, Suite 250, San Francisco, CA 94107
Phone: 415-863-9900; Fax: 415-863-9950; info@nostarch.com; www.nostarch.com

Library of Congress Cataloging-in-Publication Data

Nagata, Joe.
    [Joe Nagata no LEGO MindStorms supa kuricha. English]
    Joe Nagata's LEGO MINDSTORMS idea book / Joe Nagata ; translated by Arnie Rusoff.
          p. cm.
    Includes bibliographical references and index.
    ISBN 1-886411-40-9 (pbk.)
    1. Robots--Design and construction--Popular works. 2. LEGO toys. 3. Title: LEGO MINDSTORMS idea book. II. Title.
    TJ211.15 .N34 2000
    629.8'92--dc21
                                              00-060065

TABLE OF CONTENTS

PART 3: ADVANCED

ACKNOWLEDGMENTS
FROM THE JAPANESE EDITION

Many people helped me in producing this book. I'd particularly like to thank Mr. Jin Sato who introduced me to the world of LEGO MINDSTORMS, all the members of the editorial staff of Ohmsha's ROBOCON Magazine who encouraged me to publish this book, and everyone in the Marketing Department of LEGO Japan, Ltd.

—JOE NAGATA

THE INVENTION OF LEGO BRICKS . . .

. . . enabled us to create any shape we wanted by combining LEGO bricks.

WITH THE ARRIVAL OF LEGO TECHNIC SETS. . .

. . . we could drive gears and use motors to create complex mechanisms. And now,

LEGO MINDSTORMS SETS . . .

. . . allows us to create programs that send instructions to models
that we assemble using LEGO bricks.

This book presents a collection of unique robot creations for enthusiasts of
LEGO MINDSTORMS products. The basic theory and assembly procedure is included
for every robot, so you will learn about the internal structure of each robot and how
to build it. I hope this book will help you create your own unique robots.

INTRODUCTION

LEGO MINDSTORMS sets combine LEGO bricks with computers. It was developed by the LEGO Company after many years of research at the Media Laboratory at the Massachusetts Institute of Technology (MIT). The ROBOTICS INVENTION SYSTEM (RIS) was launched first in the United States and Europe in September 1998.

The real charm of LEGO MINDSTORMS sets is that you can freely program the movements of robots that you create using LEGO bricks. Building upon the LEGO TECHNIC series—which makes it possible to create mechanisms that move by using motors, gears, pulleys, and rubber belts—the MINDSTORMS set adds a brain, thus greatly expanding the LEGO universe.

LEGO MINDSTORMS SETS

The following is a rundown of the different MINDSTORMS sets available from LEGO.

BASIC SET (ROBOTICS INVENTION SYSTEM)

The ROBOTICS INVENTION SYSTEM (RIS) was the first of the LEGO MIND-STORMS products. It consists of a control unit, the RCX, which has a built-in computer; various types of sensors; motors; and 727 LEGO bricks.

THE RCX

The RCX (Figure 1) has a built-in 8-bit microcomputer that can control three inputs and three outputs. Light sensors or touch sensors attached to these input ports enable the robot to learn the state of the external world. Motors connected to the output ports enable the robot to move, change direction, or pick up objects.

The RCX also has an infrared (IR) port for communicating with other PCs or other RCXs, switches for operations, and a LCD panel.

NOTE *The robots described in this book use the yellow RCX from the ROBOTICS INVENTION SYSTEM. However, because most projects do not use all the input and output ports, the RCX can be replaced by the blue SCOUT from the ROBOTICS DISCOVERY SET.*

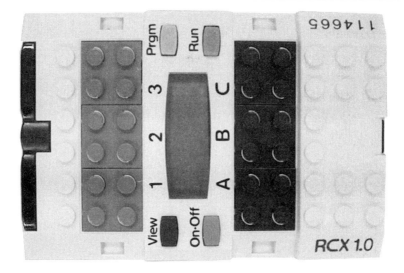

Figure 1: *The RCX*

The program code for moving the robot is written on a Windows PC and then transferred to the RCX. The PC programming environment is a proprietary GUI-based graphical programming environment that enables you to arrange commands on the screen by dragging and dropping.

ROBO SPORTS

ROBO SPORTS is an expansion set for accomplishing missions such as grabbing balls, slapping pucks, or shooting goals. The set includes an additional motor; balls; pads used in sports; various types of bricks; and a CD-ROM with 12 new missions that simulate sports.

EXTREME CREATURES

EXTREME CREATURES is an expansion set for fulfilling missions that require you to build creatures that perform various actions. In addition to various types of bricks, the set includes parts for ornamentation (such as the fiber unit which combines fiber optic strands and LEDs), and a CD-ROM with 12 new missions that simulate a variety of creatures.

EXPLORATION MARS

EXPLORATION MARS is an expansion set for fulfilling missions that require robots to be built for exploring Mars. The set includes additional tires and other types of LEGO bricks.

ROBOTICS DISCOVERY SET

This product contains a blue control unit called the SCOUT, which can be programmed independently without using a PC. The SCOUT has a built-in light sensor and can control two external touch sensors and two motors. This set enables you to easily enter the world of LEGO MINDSTORMS robots.

DROID DEVELOPER KIT

This product contains a white control unit called the MICRO-SCOUT, which has seven built-in programs. The MICRO-SCOUT has a built-in light sensor and motor. A major feature of this kit is that it enables you to build various Star Wars characters such as R2-D2.

DARK SIDE DEVELOPER KIT

This kit comes with four MicroScouts and assorted pieces that let you build Star Wars robots. A good companion to the DROID DEVELOPER KIT.

VISION COMMAND

VISION COMMAND comes with a LEGO Cam that records up to 30 frames per second and allows you to build robots that respond to what they see: motion, color, and light changes. The set also comes with the a CD-ROM with vision recognition software.

ULTIMATE ACCESSORY SET

This set will enhance your sensor collection: it contains a touch and rotation sensor, a lamp element, and a remote control, along with various other pieces. A good set if you already have a lot of basic pieces but want some more specialized robotics pieces.

IS THIS BOOK FOR ME?

If you have always wanted to make your LEGO creations come alive, or if you are simply interested in robotics, then LEGO MINDSTORMS sets and the projects detailed in this book are for you. After you've built the models in this book, you'll be ready to apply your knowledge to your own robotic creations.

The descriptions in the user's guides accompanying LEGO MINDSTORMS products contain only the minimum required information. They simply assign various missions, such as "build a robot that can trace along a black line" or "build a robot that can find and grab a ball," and no answers are provided for questions posed in the hints. They encourage you to build the robot yourself through repeated trial and error.

Once you have successfully completed the various missions detailed in the LEGO MINDSTORMS guide, you are ready to plunge headlong into the realm of creating your own unique robots. But how do you begin? This book will help you create your own fantasy robots by illustrating fundamental robotics principles with LEGO MINDSTORMS products. Building robots to suit your own fancy and having them move in whatever way you want is the true attraction and challenge of LEGO MINDSTORMS products.

WHAT YOU NEED TO MAKE THE ROBOTS IN THIS BOOK

The best basic starting set is the ROBOTICS INVENTION SYSTEM. However, most robots introduced in this book use some LEGO bricks not found in the ROBOTICS INVENTION SYSTEM set. If you want to build a robot from this book and you think that you don't have enough parts, don't despair—part of the fun is in collecting all the parts you need. But you can also make do without all the parts I've used—try to discover parts from your own collection that will work just as well. Thinking about other parts you can substitute for the missing parts or devising other methods of making parts move will start you along the path towards building your own unique robots.

PARTS LIST

The facing page shows a list of all the types of parts used throughout this book (for most types of parts, only a representative part is listed). No model uses *all* the pieces. If you want to know how many pieces a certain model requires, check the instructions for that specific model, where you'll find a detailed parts list. If you don't have all the pieces on this list, it doesn't mean you can't make the robots—try experimenting with the pieces you do have, or try collecting the pieces you need.

NOTE *In the model instructions, the pieces used in each step are pictured in the previous step lying next to the already built portion of the robot. For example, the pieces you need in Step 2 of Model 1 are pictured in Step 1.*

FINDING PARTS

If you're short on parts, you can purchase accessory sets directly from LEGO (1-800-453-4652; www.lego.com) or Pitsco LEGO DACTA (www.pitsco-legodacta.com); or you might try an online auction like Ebay (www.ebay.com). Also be sure to check out the LEGO user group www.lugnet.com for links and information on all things LEGO— you might find some users wanting to trade or sell parts.

PARTS LIST

IMAGE	EXAMPLE PIECE(S)	DESCRIPTION
	Brick 1x4 2x4	This is a standard LEGO brick. Its size is represented by the number of studs arrayed its length and width.
	Beam 1x4	This has the same shape as a brick, but it has circular holes on its side, which allow pegs to be inserted axially. The length is represented by the number of studs.
	Brick with axle hole 1x2	This brick has a cross-shaped hole in its side for keeping an axle fixed.
	Brick 2x2 slanted cut	Part of this brick is cut along a slant.
	Plate 1x4 2x4	The thickness of a plate is 1/3 that of a beam or brick. Its size is represented by the number of studs arrayed along its length and width.
	Plate 1x2 with rail	This is a plate with a rail attached to one side. The motor can be held fixed by inserting these rails in the motor's channels.
	Plate 2x2 corner	Good for reinforcing corners.
	Bracket 2x2 – 2x2	This 90 degree bracket has the same thickness as a plate.
	Hinge 1x2 hinge	Use this part for creating doors or attaching pieces at a slant.
	Liftarm 1x3 1x4 1x5 1x7	The thickness of this piece is half that of a 1x2 brick. You can insert pegs or axles into the circular or cross-shaped holes in its side.

PARTS LIST *(CONT.)*

IMAGE	EXAMPLE PIECE(S)	DESCRIPTION
	**TECHNIC triangle**	This special liftarm has a triangular shape. It is often used for building a gearbox.
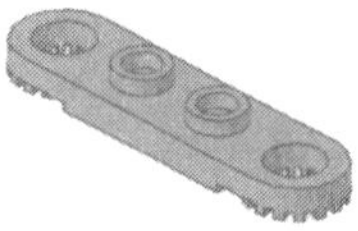	**TECHNIC plate** 1x4 with holes	Similar to a liftarm, but with grooves on each end.
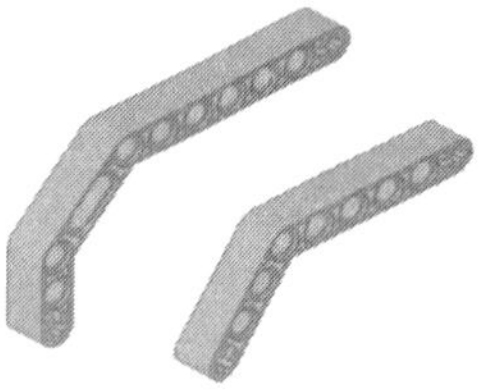	**Bent liftarm** 1x11.5 1x9	This is a brick with an angle. The bent liftarm shown on the left has 135° angles at two locations, and the one shown at the right has one angle of approximately 127°.
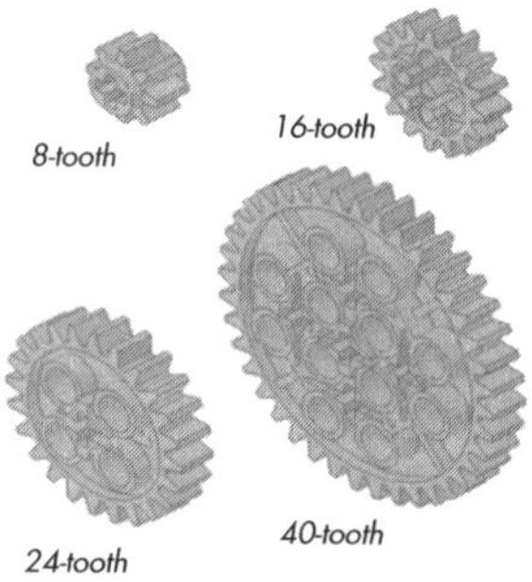 8-tooth 16-tooth 24-tooth 40-tooth	**Gears** 8-tooth 16-tooth 24-tooth 40-tooth	There are four types of flat gears with varying quantities of teeth.
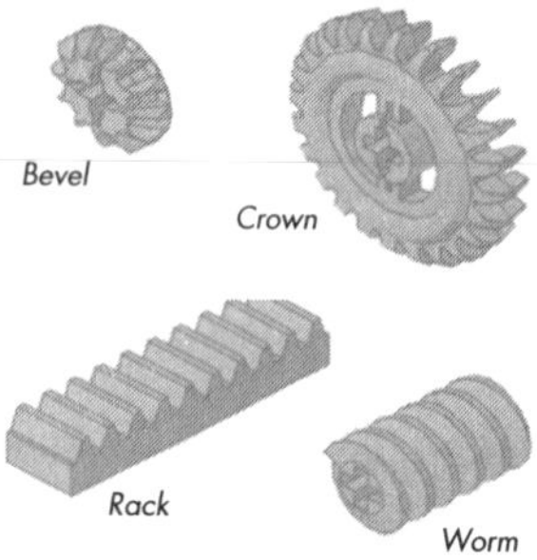 Bevel Crown Rack Worm	**Other gears** bevel crown rack worm	The specialized gears include bevel gears, crown gears, the rack, and worm gears.
	**Pulley** medium large	When attached to a rubber belt, pulleys enable the transfer of motive power.
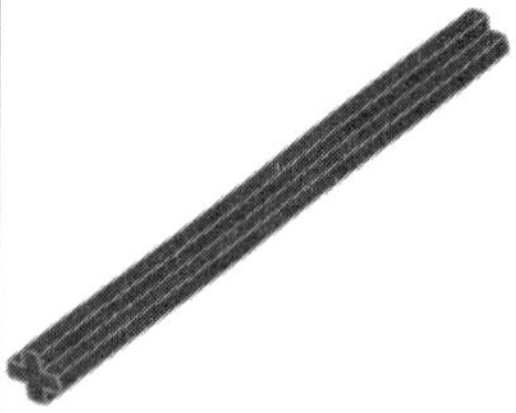	**Cross axle** 6	This component can be used as a gear axle or as a connector for bricks with holes. In the model instructions, an axle's length is abbreviated. For example, an axle of length 6 is written "an L6 axle."

PARTS LIST (CONT.)

IMAGE	EXAMPLE PIECE(S)	DESCRIPTION
	Axle joiner	This joint connects one axle to another.
	Perpendicular axle joiner 1x2 1x3	A circular hole and cross-shaped hole are oriented in perpendicular to each other. These two types of axle joiners have lengths corresponding to 2 studs and 3 studs.
	Friction pin 2 3	To use this black peg, insert it into a hole in a brick where it will be held fixed. The two types of friction pins have lengths corresponding to 2 studs and 3 studs.
	TECHNIC pin 1/2 3/4 1	Insert this gray peg into a hole in a beam where it can rotate freely.
	TECHNIC axle pin	One side of this gray peg is an axle.
	Half bushing	This part is inserted onto an axle where it is held fixed. It is half the size of a bushing.
	Bushing	This part is inserted onto an axle where it is held fixed.
	Angle connector #1 #3 #5	The angle joint lets you connect one axle to another at a fixed angle. Each number denotes a different angle.
	Motor	This is a 9 V DC motor with a connecting lead attached to the black contacts on the top.
	Connecting lead short	This cable connects the RCX to the motor or sensors.

PARTS LIST (CONT.)

IMAGE	EXAMPLE PIECE(S)	DESCRIPTION
	Light sensor	This sensor measures brightness. It emits red light from a light emitting diode (LED) and measures the brightness of the reflected light.
	Touch sensor	This sensor recognizes whether the pushbutton has been pressed.
	Fiber optic element	Place a light emitting diode (LED) in the center of this unit and when the unit rotates, light will be emitted from one of the eight holes.
	Fiber	This cable attaches to the light unit for transmitting light.
	Wheel hub small large	Wheel hub has a hole in its center for an axle. A rubber tire can be attached around its outer rim.
	Rubber tire small large	The rubber tire can be attached to a wheel hub.
	RCX	The brains of your robotic creations.
	T connector	This connector divides the flow of compressed air.
	Pneumatic tubing 15 cm	This rubber tubing, when connected to the pneumatic cylinder, pneumatic valve, or air tank, distributes compressed air.
	Pneumatic valve	This valve switches the flow of compressed air.
	Pneumatic cylinder small large	This cylinder can be lengthened or shortened by the force of compressed air.

PARTS LIST (CONT.)

IMAGE	EXAMPLE PIECE(S)	DESCRIPTION
	Air tank	This tank collects compressed air.
	Rubber band	This band can be connected to a pulley to transmit a force.
	Flexible tube	This hollow tube can be easily bent.
Flame *Head*	**SLIZER flame and head**	Decoration SLIZER parts.
Bearing *Axle*	**SLIZER bearing and axle**	SLIZER pieces for making legs.

THE PROJECTS

Here is a quick rundown of the projects in this book, categorized by level of difficulty (see the color insert for images of each project or check out the short movies of Models 6 and 7 online at http://www.nostarch.com/?robotics). Use these projects as ideas for the myriad other things you can do with MINDSTORMS sets and the types of mechanisms you can create. By all means, feel free to add sensors or motors to create your own unique robots based on the robots in this book.

PART 1: BEGINNER

MODEL 1 ROLLING CAR

The Rolling Car is a robot that will not topple over even when it crashes. The first program that I created for it made it do somersaults in place without moving forward; making the most of this "failure," I incorporated it as a skill.

The Rolling Car is appropriate even for beginners, because it can easily be built using only Robotics Invention System (RIS) parts. But don't give it short shrift; this is a very adaptable robot that can perform a variety of movements depending on the program.

MODEL 2 CENTIPEDE

The Centipede uses MINDSTORMS parts to build a form with undulating legs like that of a centipede. Because the insect is made of repeating modular parts, builders with only a few parts can make a short centipede and builders with lots of parts can make a longer one.

MODEL 3 WATER SKATER

The Water Skater is the result of repeated attempts to produce amusing motion using a shaft. The robot's complex motion results from a simple mechanism. Check it out yourself.

PART 2: INTERMEDIATE

MODEL 4 LEGOSAURUS

I created the LEGOsaurus using tires and LEGO TECHNIC SLIZER parts to build the legs. The torso turns out to be surprisingly compact because of the use of bent liftarms. If you want to build four-legged robots, you should first master this basic form.

MODEL 5 TRAIN

This project resulted from trying to come up with practical ways to use flexible tubes. The techniques used to create the Train can also be adapted to create a monorail. Try it!

MODEL 6 WALKER

When I first built a two-legged robot, it couldn't walk with the RCX mounted on top. I solved that problem with this second-generation Walker, which was able to walk with the mounted RCX once I made the legs and hip more sturdy.

I'll also show you how to build this Walker using only RIS parts.

PART 3: ADVANCED

MODEL 7 CLIMBER

The Climber changes its shape as a means of climbing over an obstacle. Because its actions are complex, I used NQC for the program.

MODEL 8 LEGO CLOCK

Obviously, a clock made of LEGO bricks and gears. If you wish to learn techniques for fitting gears together, try building this model.

MODEL 9 PNEUMATIC ENGINE CAR

This car applies the pneumatic system, which is a cylinder that uses compressed air. If you have this cylinder part, you should definitely try building this model.

MODEL 10 BEETLE

This Beetle is a walking robot that can turn left or right by changing the length of its stride. Although its construction is complicated, don't miss this chance to experience the essentials of walking robots.

PLANNING YOUR OWN ROBOTS

Before we dive into the projects, let's look at the steps involved in building your own unique robots.

1. PURPOSE AND THEORY

First, clarify the purpose of your robot and the theory behind it. You can adapt a standard LEGO mission or participate in one of the frequently held contests or competitions (see www.mindstorms.lego.com for contest times and locations). But whatever you do, be sure that the robot you dream up is one that you really want to build.

2. ASSEMBLY

Next, think about how to organize your robot to achieve your objective. One of the great things about MINDSTORMS sets is that you can try out all kinds of mechanisms time and again and learn what works through trial and error. As you try various ways to make something happen, you'll eventually find the best method.

3. PROGRAM

To make your robot move, you ultimately must load a program into the RCX that will control the output to motors based on sensor input. You will usually determine the best way to balance this output using trial and error to figure out how long the motor should turn when a sensor value reaches a certain level. Be sure to go through the CD tour that comes with your MINDSTORMS set so you can learn how to program the RCX.

I usually go through all three steps—Theory/Purpose, Assembly, and Program—when I create my robots. However, once in a while, an interesting structure or program idea occurs to me right away and I skip one or two steps. Nevertheless, combining these three steps with some trial and error will provide you with the shortest route to completing your own robots.

Now let's build some robots!

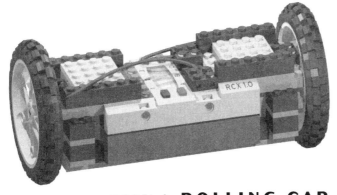

MODEL 1: **ROLLING CAR**

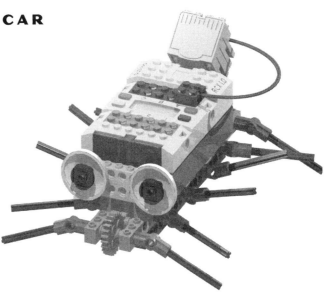

MODEL 2: **CENTIPEDE**

MODEL 3: **WATER SKATER**

PART 1
BEGINNER

MODEL 1: ROLLING CAR

One of the very first things almost everyone makes when they get their MINDSTORMS kit is a robot that consists of a motor and tires. Then, when it's allowed to run freely in a narrow room, it crashes and overturns.

This chapter will show you how to build a vehicle that won't overturn.

This robot consists of two tires and a motor. Even when it overturns, it returns to its upright position. If you look at this robot from the side (Figure 1-1), you'll see why: Its body is entirely inside the circumference of the tires, so its center of gravity is lower than the wheel axles. Therefore, when the body is turned upside down, its own weight forces it to return to its original upright position.

Figure 1-1: *Photograph of the Rolling Car viewed from the side*

Although this robot has a simple structure, it can perform a variety of movements, depending on how you program it. For example, not only can it move forward, backward, left, and right, but it can also stay in one place and do somersaults without moving its tires. (See "Doing Somersaults," below.)

PARTS LIST

TYPE	SIZE	QUANTITY
Brick	1x6	6
Brick	2x2	8
Plate	1x1	4
Plate	1x2	10
Plate	1x2 with rail	8
Plate	1x4	8
Plate	2x4	9
Plate	2x6	2
Plate	2x8	5
Plate	2x10	3
Motor		2
Connecting lead	Short	2
Wheel hub	Extra large	2
Rubber tire	Extra large	2

BUILD THE MOTOR HOUSING

1

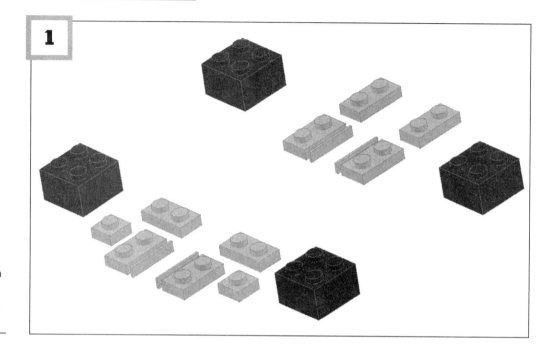

Collect the parts shown in the figure for the frame that supports the motor.

2

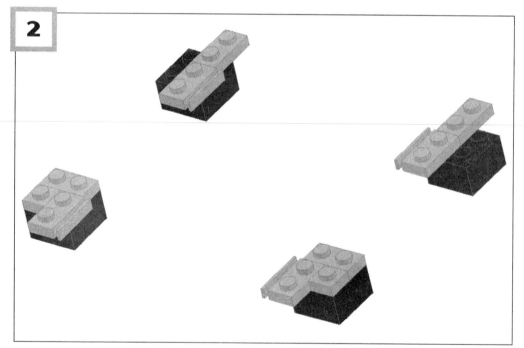

Attach the 1x2 plates with rails to the 2x2 bricks as shown. Attach the other 1x2 and 1x1 plates.

3

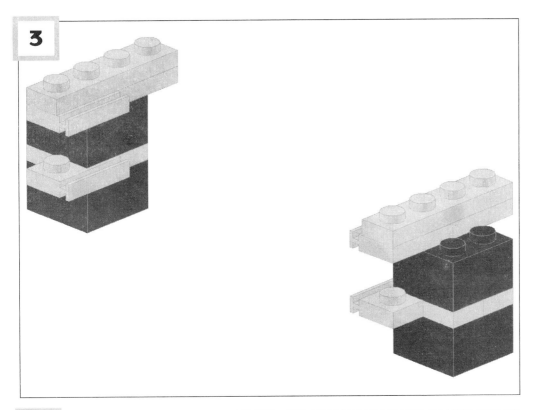

When the assembled bricks are stacked as shown, two rails are produced that will align with the motor's channels (the grooves in the side of the motor).

4

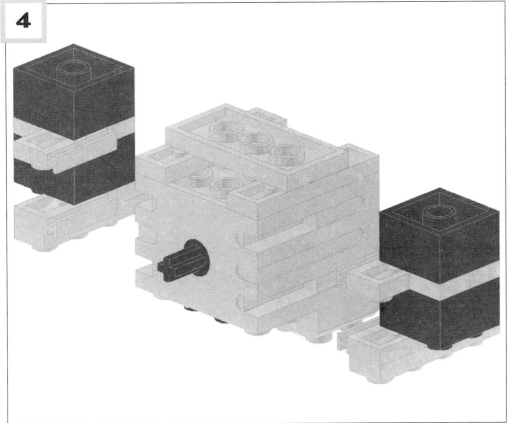

Invert the motor and the two brick assemblies. Attach two 1x4 plates to the bottom of the motor, then connect them with a 2x4 plate.

5

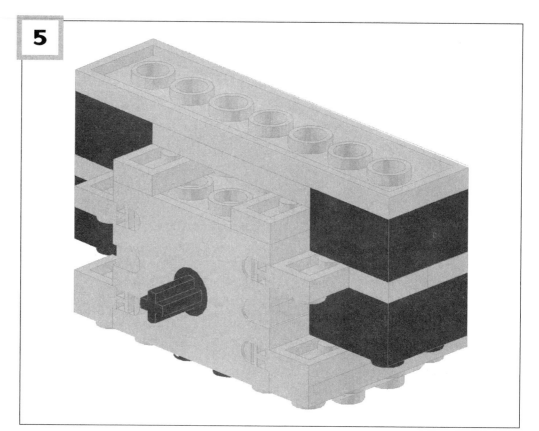

Align the parts that you built in Step 3 so you can insert the rails in the channels on both sides of the motor, then attach a 2x8 plate to secure the entire assembly.

6

Further reinforce the assembly with a 2x6 plate and attach a 1x6 brick as shown in the figure to complete your first of two motor housing assemblies. Now repeat Steps 1 through 5 to build one more motor housing, exactly like the one you've just finished.

7

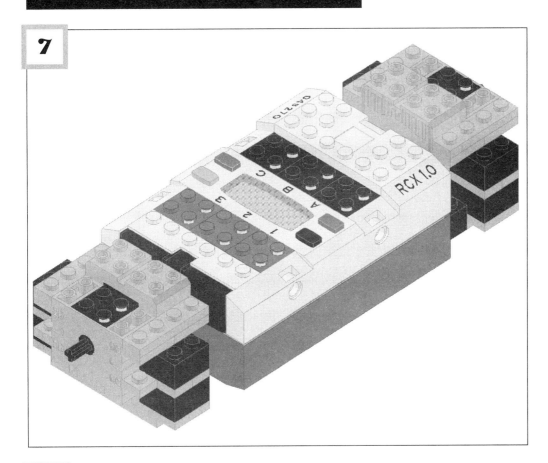

Align the two motor housing assemblies with the front and back of the RCX, as shown in the figure.

8

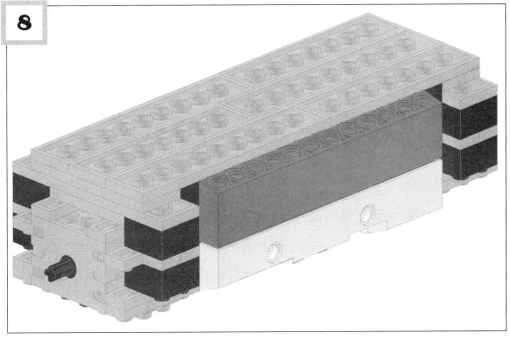

Fasten the bottom of the RCX into position. Use three 2x10 and three 2x8 plates to hold the RCX and motors together.

The height of the RCX differs at the front and back; to correct this, attach two stacked 2x4 plates on the left side of the RCX then attach one 2x4 plate on its right side, as shown in the figure.

Also attach a 1x4 plate to each side of the left-hand motor. Then, stack one 1x4 and two 1x2 plates and attach them to both sides of the right-hand motor to make the height of the motor housing assemblies equal to the height of the RCX.

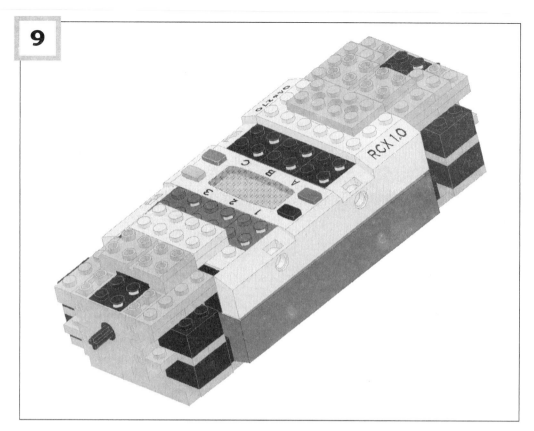

Attach two 2x4 plates to the left side and two to the right side of the RCX to join the RCX and motors, then further reinforce them using four 1x6 bricks as shown in the figure.

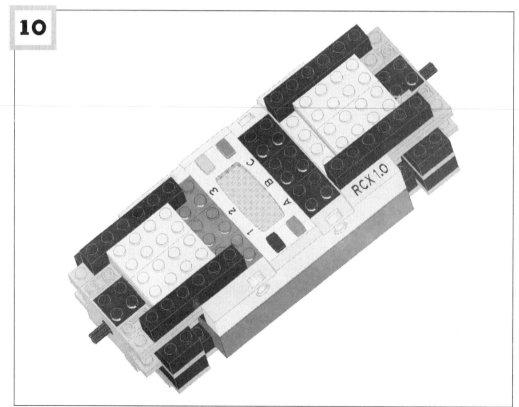

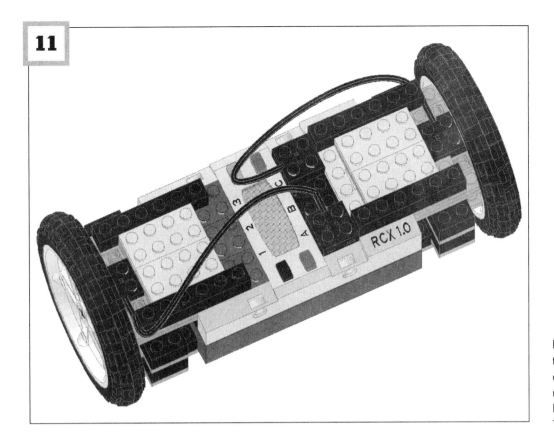

11

Finally, attach large tires to the motor shafts and wire the RCX to the motors to complete the Rolling Car.

NOTE *If you start your robot now, it will spin wildly because the motor starts too quickly. To fix this problem, use the program below to gradually increase the motor's power to produce a smooth acceleration.*

PROGRAM

The Rolling Car is really just a simple robot, but we can make it do interesting things by controlling the left and right wheels in various ways. To make the car

- *Move forward*, rotate the left and right wheels in the same direction.
- *Turn right*, stop the right wheel and rotate the left wheel.
- *Turn left*, stop the left wheel and rotate the right wheel.
- *Rotate*, rotate the left and right wheels in opposite directions.

DOING SOMERSAULTS

Let's make the body do a somersault without moving the wheels. The question is how to set the controls to achieve the difference in two movements, with the motors attached to the left and right wheels both rotating in the forward direction. The answer is to use inertia and control the amount of power sent to the motor.

Ordinarily, to move the car forward steadily, we begin by sending the lowest amount of power to the motor (specified by Set Power in the program) and gradually increasing it. However, to do a somersault in place, we rotate the motor suddenly with maximum force. The sample program below (Figure 1-2) moves the robot in an elegant dance that has a grand finale.

What kind of movements will you make it do?

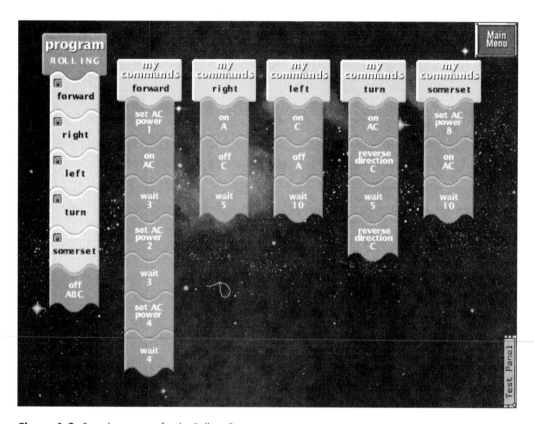

Figure 1-2: Sample program for the Rolling Car

TROUBLESHOOTING

• If your robot's wheels turn in opposite directions, turn one motor's connecting lead 180 degrees on the RCX. If this doesn't fix the problem, review the Guided Tour on the MINDSTORMS CD to familiarize yourself with wiring the motor.

• If your wheels rub against the side of the robot's body, try adding an axle joiner and cross axle to lengthen the axle.

TIP

GIANT TIRES

LEGO MINDSTORMS kits provide a variety of tires. You can expand on this variety by combining them.

For example, you can take large tires in the RIS set and turn them into giant tires. To do so, place two of these large tires side by side, then place a caterpillar tread around them. Because the caterpillar tread is the same size as the tire's outer circumference, the combination produces one giant tire. You could place two caterpillar treads around three tires to make an even bigger tire. Why not turn one of these wider tires into a unicycle?

MODEL 2: CENTIPEDE

In this project, we'll build a robot that walks like a centipede. But first, to better imagine how to approach this project, think about how a six-legged insect-like creature moves forward. The legs of a centipede undulate; when the front and back legs point forward, the center leg points backward, forming a kind of sine wave.

If you have enough LEGO parts, you can build a centipede of just about any length, as long as the motor has enough power to drive the legs. But rather than create a monster, let's start with a twelve-legged robot with six legs on each side, as shown here.

When a robot has lots of legs, one leg at a time pushes the robot forward, with the rest of the legs following in sequence; the robot moves forward little by little, as each leg in turn contributes to the robot's movement.

Because all the legs are moving at once, it's important to prevent legs from bumping into each other. To do so, we offset the relative orientation of adjacent legs a little at a time. The overall result is that the legs appear to undulate as in Figure 2-1.

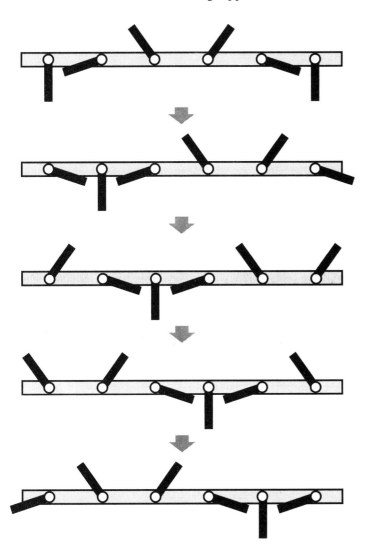

Figure 2-1: *Movement of Centipede legs*

NOTE *Instead of attaching the legs as shown in Figure 2-1, you can reverse their orientation 180 degrees (look at the picture with the book turned upside down). Why not try each type of movement?*

PARTS LIST

TYPE	SIZE	QUANTITY
Brick	2x4	2
Beam	1x6	2
Beam	1x8	2
Beam	1x16	6
Hinge	1x2	2
TECHNIC triangle		2
Gear	8-tooth	5
Gear	24-tooth	6
Gear	Worm gear	1
Cross axle	2	2
Cross axle	4	5
Cross axle	6	19
Axle joiner		1
Perpendicular axle joiner	1x2	2
Half bushing		22
Angle connector	#3	12
Motor		1
Connecting lead	Short	1

ASSEMBLY PROCEDURE

1

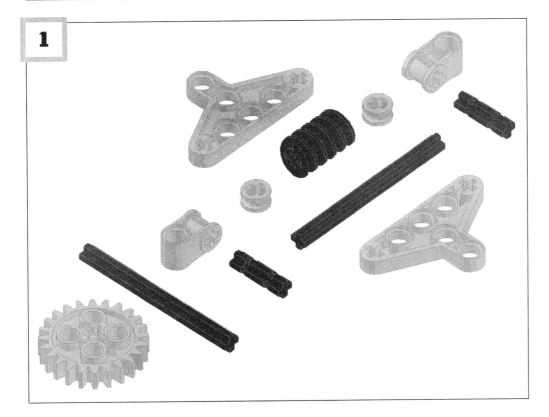

Gather the parts for the worm gear assembly.

2

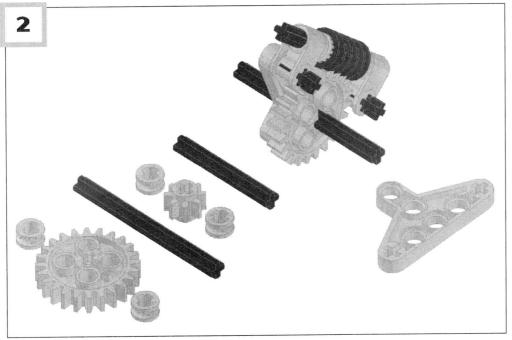

Use L2 cross axles to fasten perpendicular axle joiners to a TECH-NIC triangle. Pass an L6 cross axle through a 24-tooth gear and attach it between the triangles. Pass an L6 cross axle through the perpendicular axle joiners, attaching the worm gear and two half bushings as shown.

3

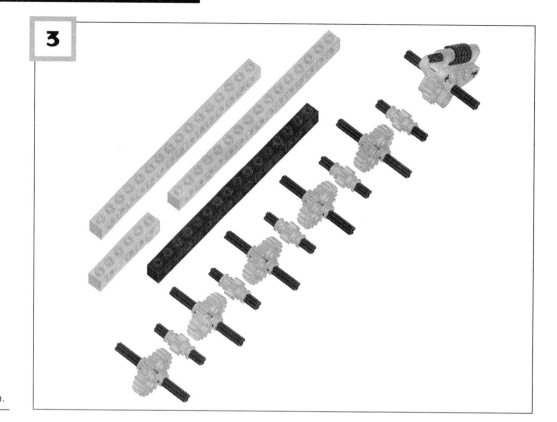

Pass L6 cross axles through five 24-tooth gears and L4 cross axles through five 8-tooth gears; attach half bushings on both sides of each gear. Lay them out as shown.

4

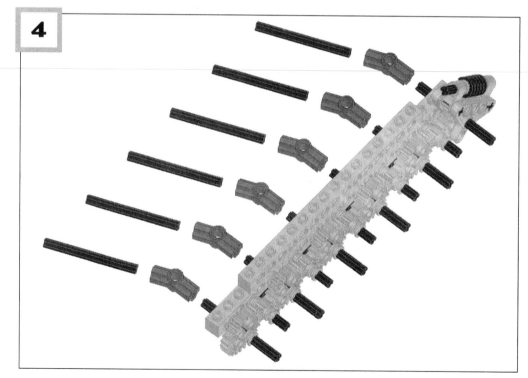

Pass the cross axles through a 1x6 and a 1x16 beam, and use 1x16 beams to reinforce the top and bottom of those beams. Gather parts for the legs.

5

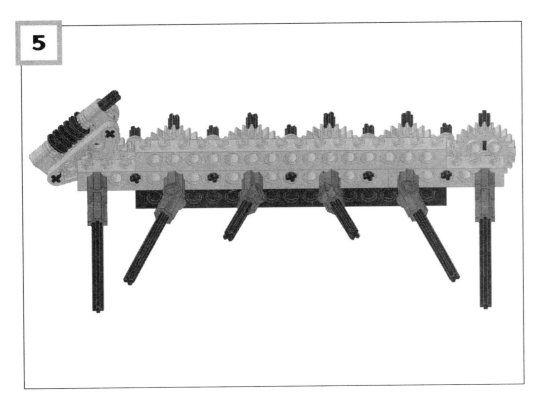

Place a #3 angle connector on each L6 cross axle protruding from the 24-tooth gears, and attach a second L6 cross axle to each #3 angle connector to create the legs. Adjust the meshing of the gears so that the legs, when viewed from the side, are each offset just a little from its neighbor, as shown. If you have trouble adjusting the legs, either disconnect and reattach them at the desired positions or remove the 8-tooth gears, adjust the legs, and reinsert the 8-tooth gears with the legs in the desired positions.

6

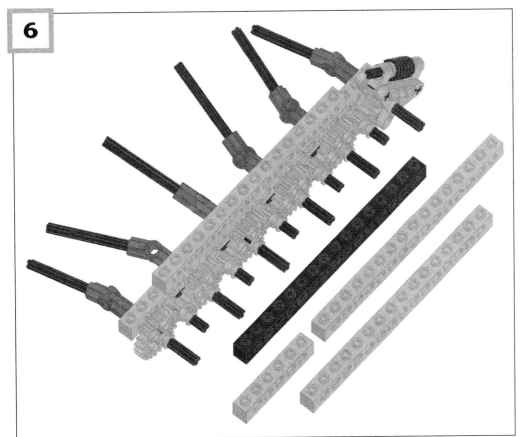

Before continuing, turn one of the gears to make sure that none of the legs hit each other, then collect the beams for the other side as shown.

7

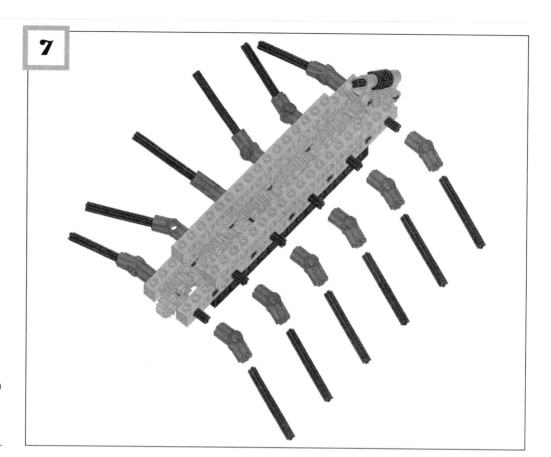

Combine the 1x6 and 1x16 beams and attach them to the torso as in Step 3, then collect the parts for the legs.

8

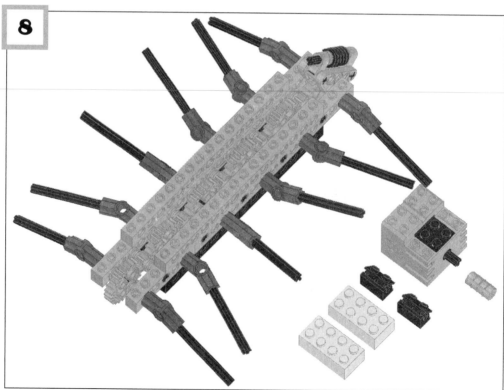

Attach the #3 angle connectors and the L6 cross axles to the axles as in Step 5. Line up each new leg so its movements will match the leg on the opposite side, as shown. Collect the bricks and hinges to support the motor.

9

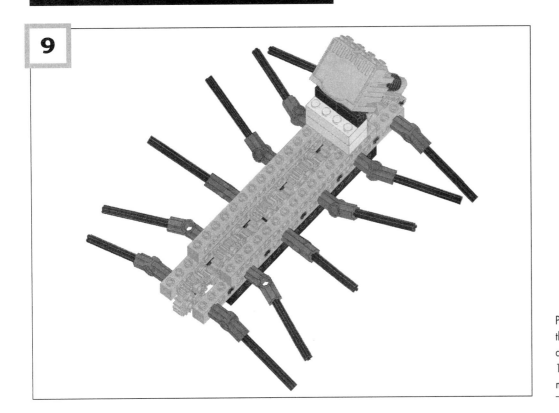

Place an axle joiner on the motor cross axle and attach 2x4 bricks and 1x2 hinges to secure the motor onto the torso.

10

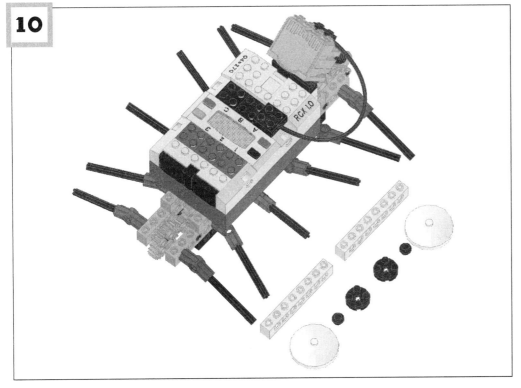

Mount the RCX on the body and connect the motor to output A with the connecting lead. This completes the robot. Collect the parts for the "eyes."

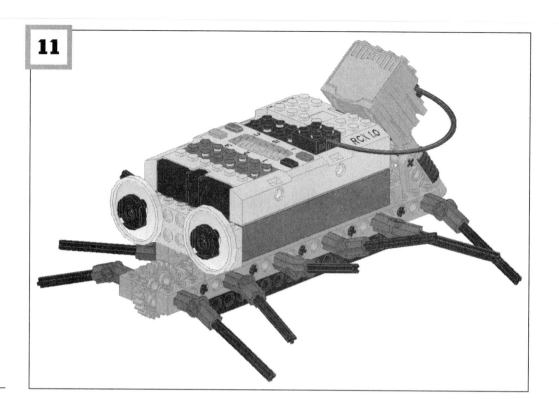

11

Attach eyes to the front so that the robot appears as shown.

PROGRAM

This robot, which uses only output A, can move according to the default program set in the RCX. Use the "Prog" button to select Program 1 and press the "Run" button to start the program.

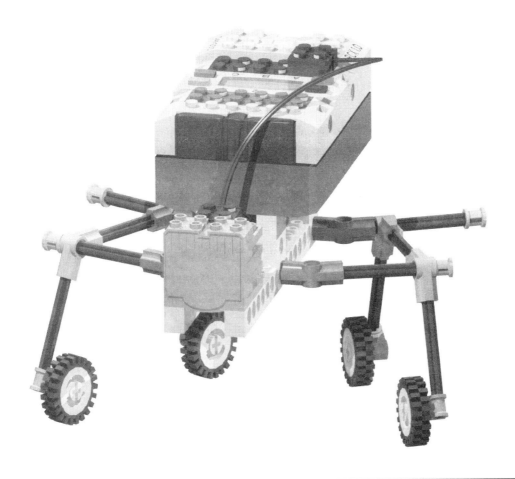

MODEL 3: WATER SKATER

We usually propel a robot by rotating its tires or moving its legs. In this chapter we introduce a unique, unconventional method of propulsion.

The four tires in this robot can only coast because they are not connected to the motor. Further, the axles attached to the angle connectors are not fixed and rotate as the motor turns the gear assembly. So how does this robot move? It skates.

Think about how we move when ice skating, but not about sliding one foot at a time. While our robot will skate, it does so a bit differently than what we may be used to: It slowly moves forward by repeatedly forming a V shape with both feet, gradually letting its feet move farther apart, and then reversing the angle of both feet so that they come together again (see Figure 3-1).

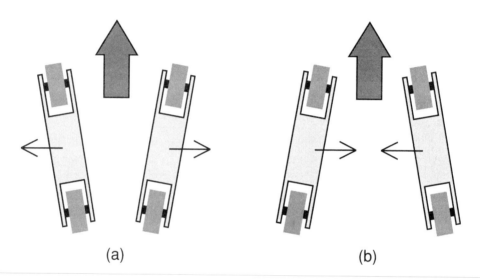

(a)   (b)

Figure 3-1: *How the Skater moves*

Were you to try this movement on ice skates, you would need to repeatedly force your feet together, then apart, then together. However, because this robot is not moving on ice, it relies on coasting tires, which rotate forward or backward but not left or right. The feet diverge and converge using the unique method shown in the following photographs (Figure 3-2) of the Water Skater in action.

As you can see, this extremely simple structure produces a very complex kind of movement.

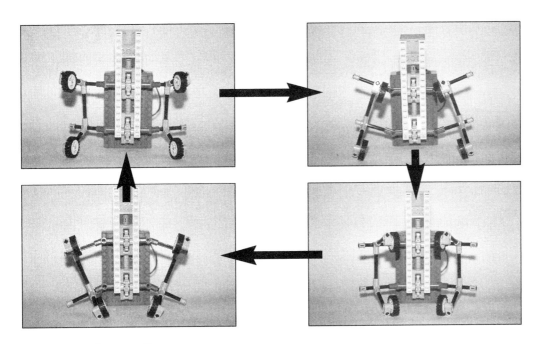

Figure 3-2: *The Water Skater in motion*

PARTS LIST

TYPE	SIZE	QUANTITY
Beam	1x2	8
Beam	1x16	2
TECHNIC triangle		4
Gear	24-tooth	2
Worm gear		2
Cross axle	2	4
Cross axle	6	14
Axle joiner		1
Perpendicular axle joiner	1x2	8
Half bushing		5
Bushing		5
Angle connector	#3	4
Angle connector	#5	4
Motor		1
Connecting lead	Short	1
Wheel hub	Small	4
Rubber tire	Small	4

BUILD THE GEAR ASSEMBLY

1

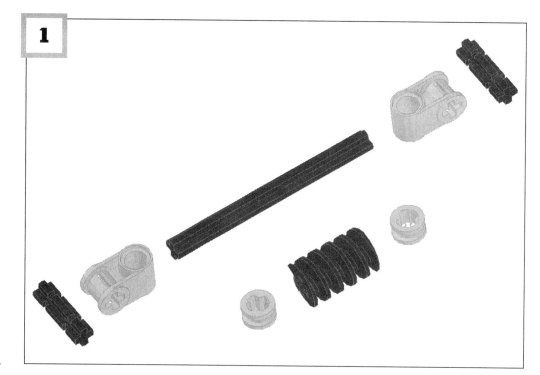

Since moving this robot requires a powerful torque (force for rotating the axle), we'll use a worm gear. The remaining pieces in the figure form the basis of the gear assembly.

2

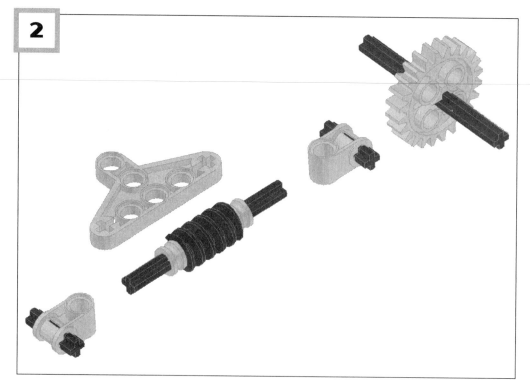

Slide the worm gear onto an L6 cross axle and place half bushings at both ends to secure the gear as shown. Insert L2 cross axles into two perpendicular axle joiners, then collect the pieces for the next step.

3

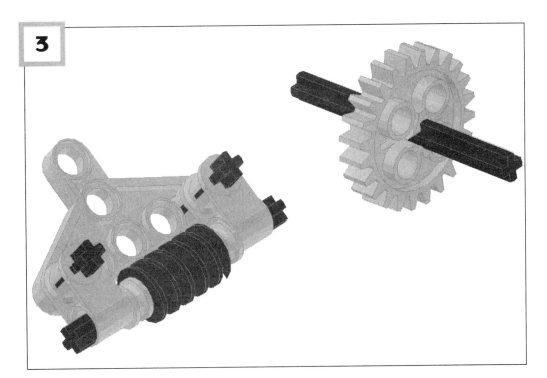

Pass the worm gear axle through the two perpendicular axle joiners, then attach the unit to a TECHNIC triangle. Pass a cross axle through the 24-tooth gear.

4

Mount the axle and gear onto the TECHNIC triangle so that the teeth align with the worm gear.

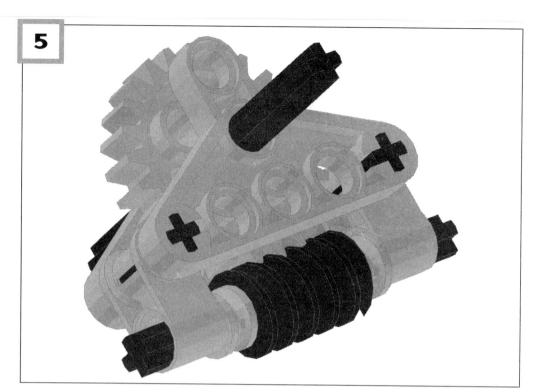

Fix the 24-tooth gear in place by attaching another TECHNIC triangle as shown.

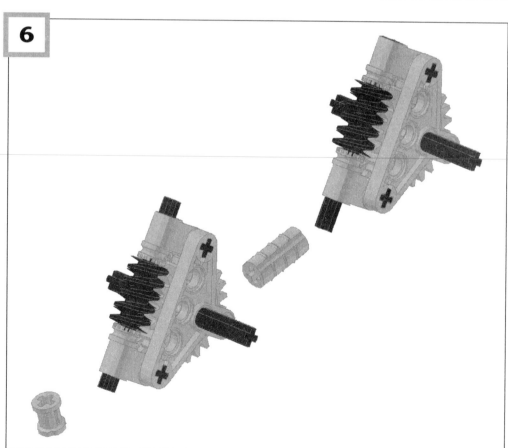

6

Follow Steps 1 through 5 above to build a second, identical worm gear unit.

7

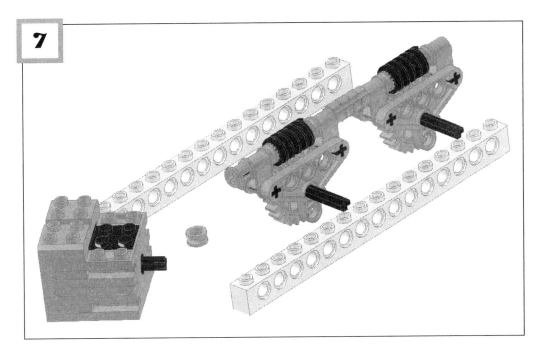

Use an axle joiner to connect the cross axles in the two worm gears. Align the cross axles so that their orientations match. When properly aligned, the axles' crosses should have the same orientation when viewed end-on (see NOTE below).

Attach a bushing to the end of the axle that will attach to the motor.

NOTE *To see if axles are properly aligned look at them head-on and check that their crosses are pointed in the same direction. If we represent the axles' crosses as plus signs, then correctly aligned axles would look like this:*

✚   ✚

Incorrectly aligned axles would look like this:

✚   ✖

8

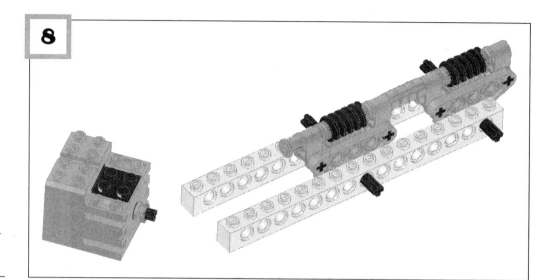

Slide two 1x16 beams onto the cross axles, then attach a half bushing as far as possible onto the motor shaft.

9

Snap the motor in place on top of the two 1x16 beams while inserting the motor shaft into the bushing, connecting it to the worm gear axle.

10

Extend L6 cross axles outward with #3 angle connectors, orienting all four connectors in the same direction. Next, assemble eight 1x2 beams (or bricks) as shown.

11

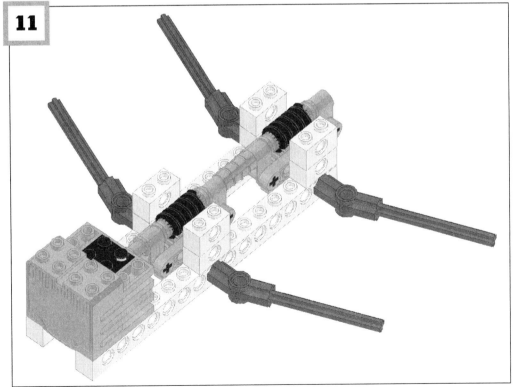

Attach the 1x2 beams (or bricks) as shown to form a base for mounting the RCX.

12

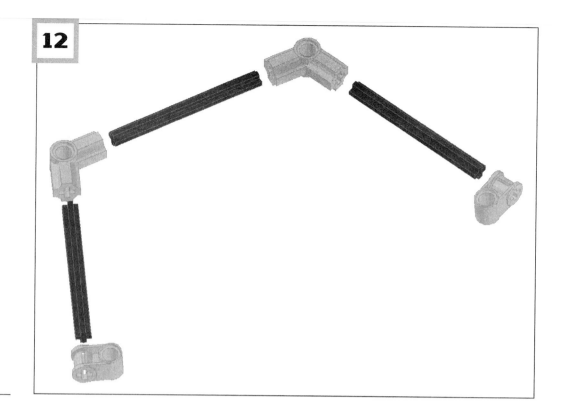

Align three L6 cross axles, two #5 angle connectors, and two perpendicular axle joiners as shown to build a foot.

13

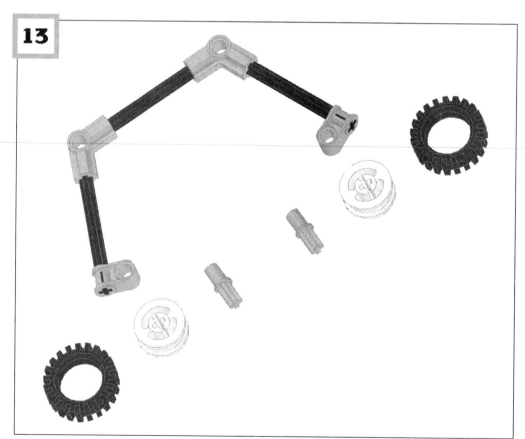

Begin building the foot by connecting the three cross axles with the #5 angle connectors as shown. Lay out tires, wheels, and gray TECH-NIC pins as shown.

14

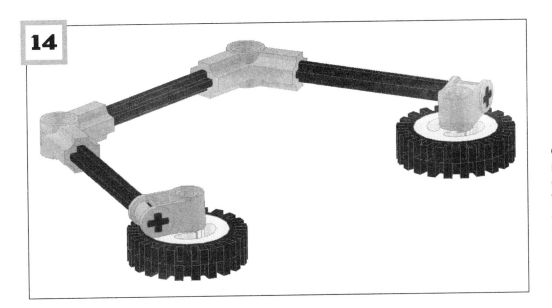

Complete the foot by inserting the wheels into the tires and the gray TECHNIC pins into each axle so that the tires turn freely. Next, build one more identical foot, and assemble both feet as shown.

ATTACH THE FEET AND RCX TO THE BODY

15

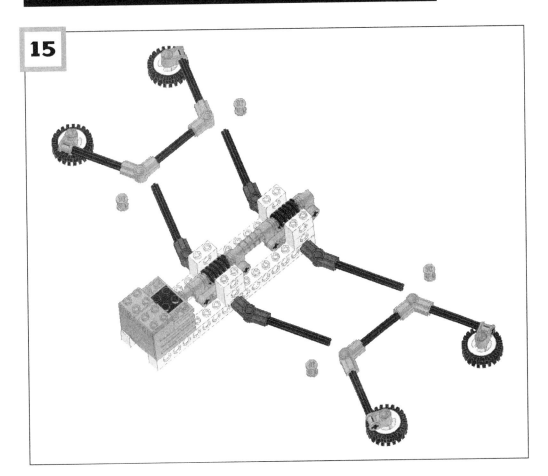

Slide the two feet onto the axles extending from the torso by sliding the axles through the angle connectors on each foot. When properly connected, the assembled wheels should be oriented as shown.

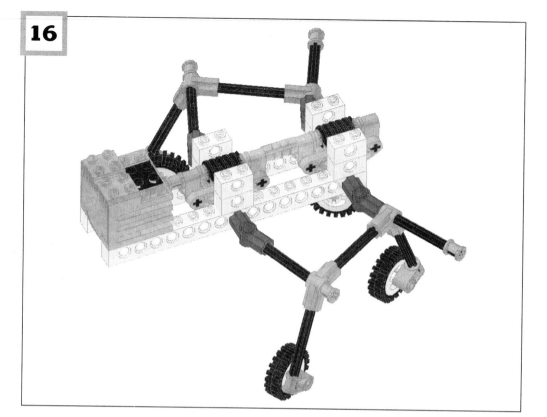

Snap bushings onto the ends of the axles extending from the torso to secure the feet.

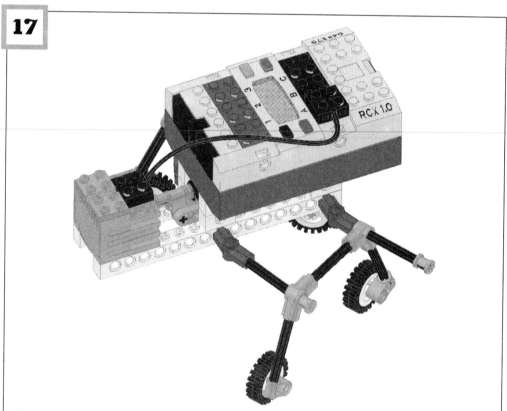

Mount the RCX to the 1x2 beams (added in Step 11), then link the motor to output A of the RCX with a short connecting lead. Your Water Skater is now complete.

PROGRAM

This robot uses only output A and can move according to program 1, set by default in the RCX. Use the "Prog" button to select "1" and press the "Run" button to start the program.

MODEL 4: **LEGOSAURUS**

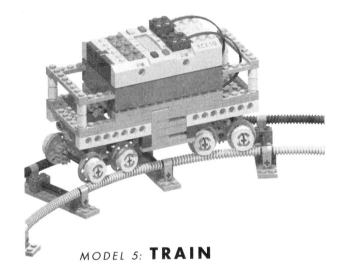

MODEL 5: **TRAIN**

MODEL 6: **WALKER**

PART 2
INTERMEDIATE

MODEL 4: LEGOSAURUS

In this project, we'll create a dinosaur-like robot that lumbers on four legs. The LEGOsaurus propels itself forward with a motor that slowly moves both front and back legs. As a final artistic touch, we'll give our robot some character with LEGO TECHNIC SLIZER parts for the head and legs.

The primary challenge we face in building a four-legged walking robot is to make the legs diagonally opposite each other move in parallel. That is, both the right front leg and rear left leg should move forward simultaneously.

Figure 4-1 shows how this movement works. Note that when the leg positions begin as shown in (a) the robot takes one step forward by lifting the right front and left rear legs (b) and moving them both forward (c) simultaneously. When the legs reach position (d) both feet return to the ground.

To help the robot balance, we'll mount the RCX so that its center of gravity (shown as the center black dot in Figure 4-1) remains centered over the grounded feet as the robot moves.

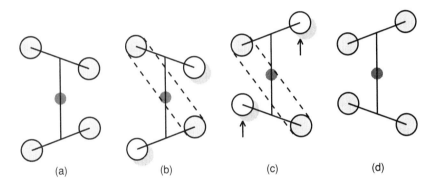

(a)   (b)   (c)   (d)

Figure 4-1: *Center of gravity when walking on four legs*

The first three photographs below (Figures 4-2 through 4-4) show the LEGOsaurus in action and correspond to positions (a) through (d) in Figure 4-1. Figure 4-5 shows the opposite foot lifted to begin another step.

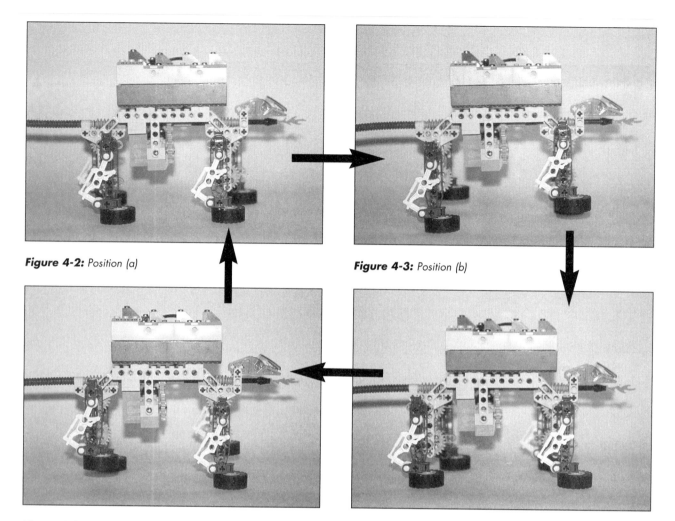

Figure 4-2: Position (a)

Figure 4-3: Position (b)

Figure 4-5: Beginning a new step

Figure 4-4: Position (d)

PARTS LIST

TYPE	SIZE	QUANTITY
Brick	1x1 round	1
Beam	1x2	2
Beam	1x10	2
Brick	2x2 slanted cut	3
Plate	1x2	6
Plate	1x2 with rail	4

PARTS LIST *(CONT.)*

TYPE	SIZE	QUANTITY
Plate	2x6	4
Liftarm	1x3	10
Liftarm	1x5	2
TECHNIC triangle		4
Bent liftarm	1x11.5	4
Gear	8-tooth	2
Gear	24-tooth	6
Gear	Worm gear	2
Cross axle	2	15
Cross axle	3	6
Cross axle	4	2
Cross axle	6	4
Cross axle	10	1
Cross axle	12	1
Axle joiner		1
Perpendicular axle joiner		4
TECHNIC pin with axle		4
Friction pin	2	4
TECHNIC pin	3/4	4
Half bushing		28
Motor		1
Connecting lead	Short	1
Wheel hub		4
Rubber tire	Medium	4
SLIZER axle		4
SLIZER bearing		8
SLIZER plate		4
SLIZER head		1
Flame		1

BUILD THE LEGS

1

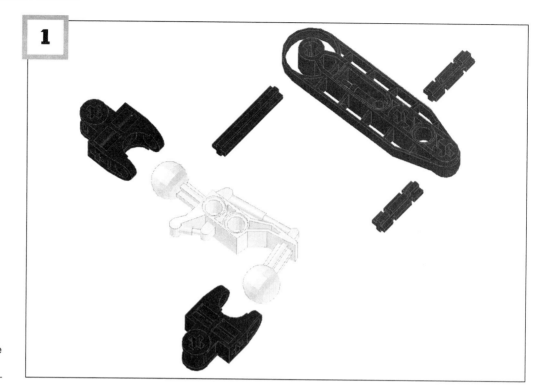

We'll use LEGO TECH-NIC SLIZER parts to build the legs. Gather three cross axles (two L2s and one L3) and the SLIZER parts shown.

2

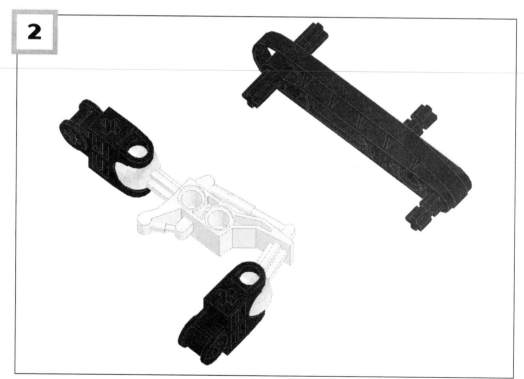

Attach two SLIZER bearings to a SLIZER axle, then insert the three cross axles through the SLIZER plate as shown.

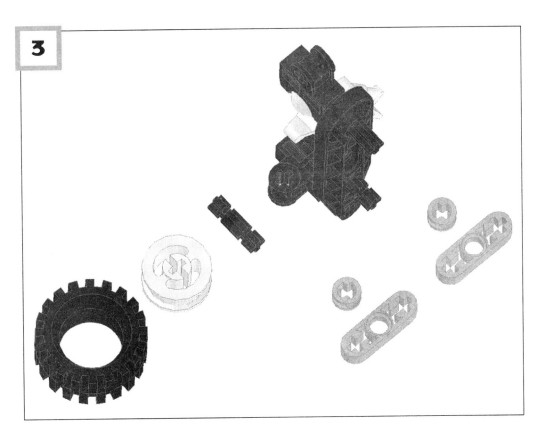

3

Connect the SLIZER bearings to the SLIZER plate. Orient the bearings as shown, then attach them to the L2 and L3 cross axles extending from the SLIZER plate. Gather the parts shown for Step 4.

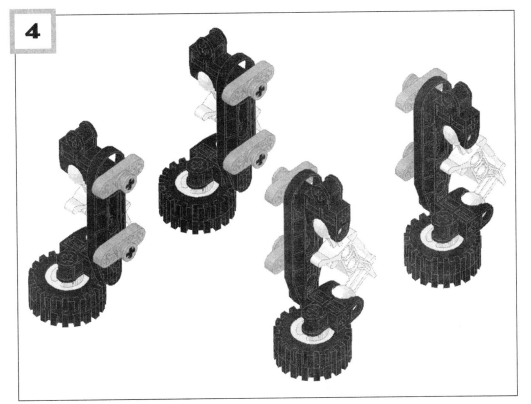

Build the foot by attaching medium tires to the leg with an L2 cross axle. Next, pass the protruding axles through the center holes in 1x3 liftarms, and use half bushings to lock the liftarms in place as shown. Build three more legs to end up with two right and two left legs.

5

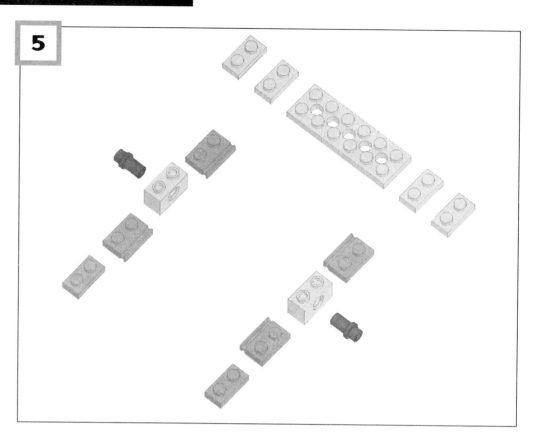

Assemble the pieces shown.

6

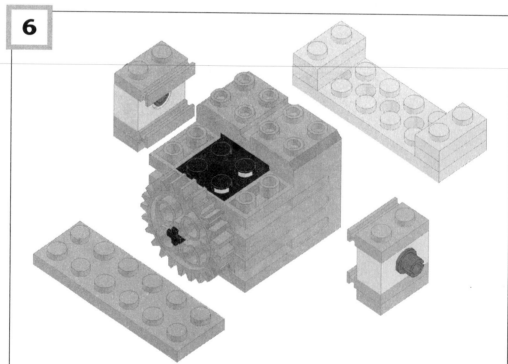

Build the various elements and assemble them as shown. Attach a 24-tooth gear to the motor and one 3/4 TECHNIC pin to each 1x2 beam.

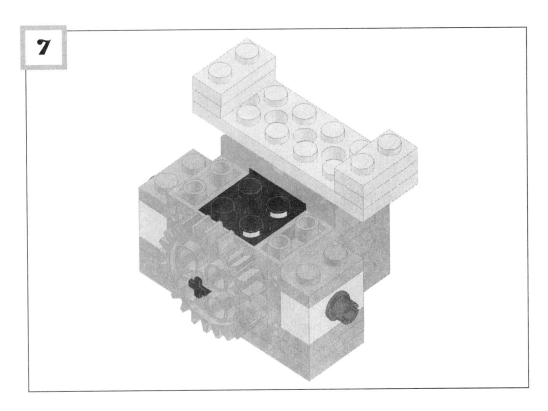

7

Align the 1x2 plates with rails with the channels on both sides of the motor and attach a 2x6 plate to the bottom to fix them in place. Use two 1x2 plates to attach the 2x6 plates to the top of the motor as shown.

BUILD THE GEAR ASSEMBLY

8

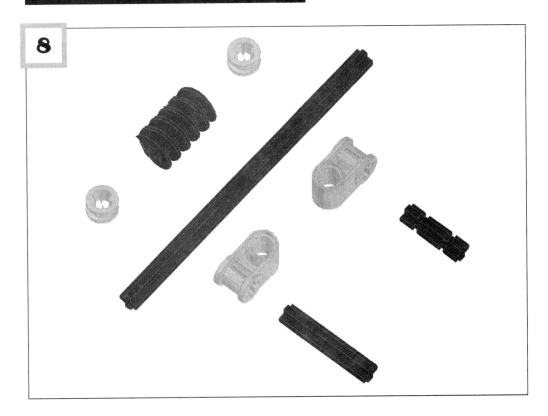

Since walking on four legs requires a lot of torque, we'll use a worm gear for the gear assembly. Gather the parts shown in the figure to build this assembly.

9

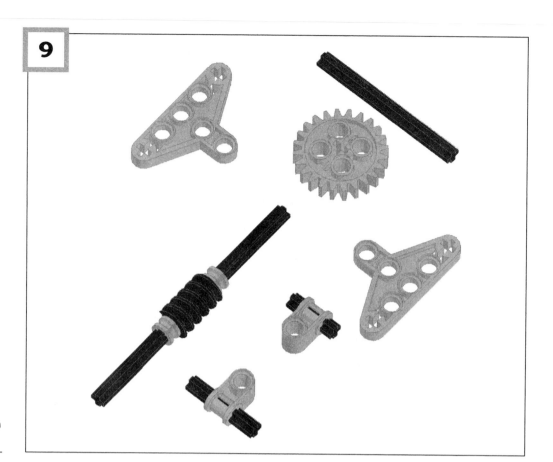

Center the worm gear on the L10 cross axle and secure it with half bushings. Gather two perpendicular axle joiners. Insert an L3 cross axle through one and an L2 axle through another.

10

Attach one perpendicular axle joiner to each end of the L10 axle. Pass an L6 cross axle through a 24-tooth gear and position the 24-tooth gear under the worm gear.

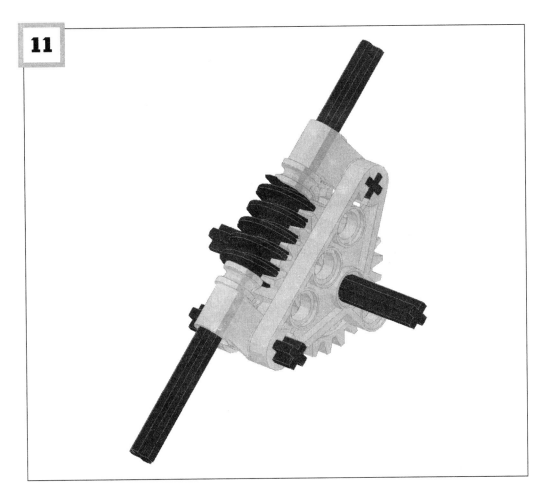

11

Enclose the 24-tooth gear with two TECHNIC triangles by passing the ends of the L6 cross axle through the center holes of each triangle. Next, pass the axles of each perpendicular axle joiner through the outermost holes of the triangles to complete the first gear assembly. (The perpendicular axle joiner with the L3 cross axle marks the front of the gear assembly.)

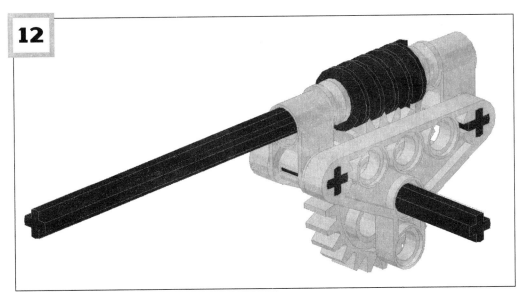

12

Follow Steps 8 through 10 above to build one more gear assembly, but use an L12 cross axle instead of an L10, and secure both perpendicular axle joiners with L2 cross axles only (rather than L2 and L3 cross axles). Position the worm gear as shown.

13

Gather the pieces shown to begin building the body.

14

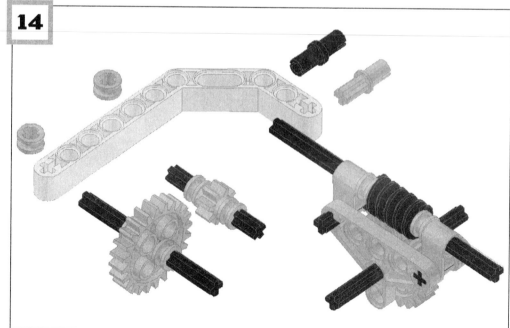

Attach a 24-tooth gear to an L6 cross axle and an 8-tooth gear to an L4 cross axle; then add half bushings to each side of the gears as shown.

15

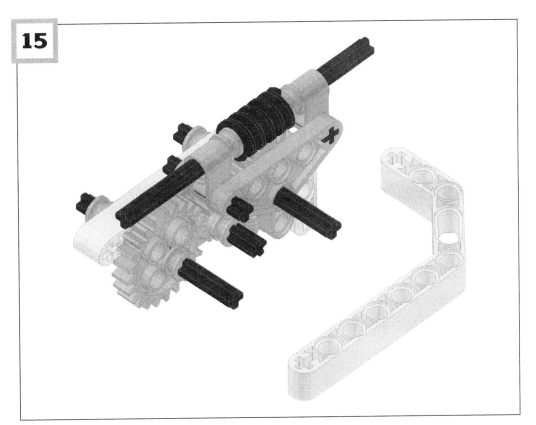

Attach the worm gear assembly and other gears to a 1x11.5 bent liftarm by inserting the 24-tooth gear through the first hole, the 8-tooth gear through the third hole, and the worm gear assembly through the fifth hole. Center the L3 cross axle above the long side of the bent liftarm, as shown, and mesh the gears so that the cross axles at the top and bottom of the assembly align when viewed from the side (see the Note after Step 7 in Model 3 for what constitutes proper alignment). If the cross axle alignment is off you will not be able to attach the legs.

16

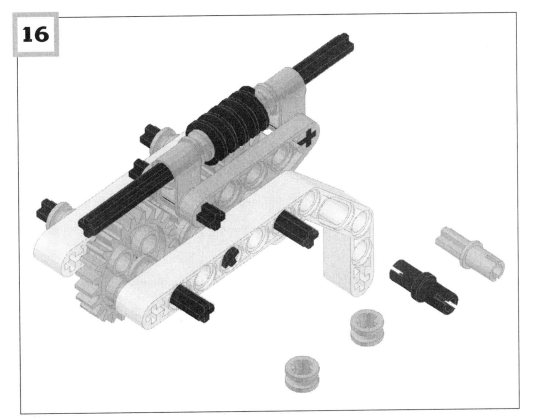

Attach a bent liftarm to the other side of the gear assembly.

17

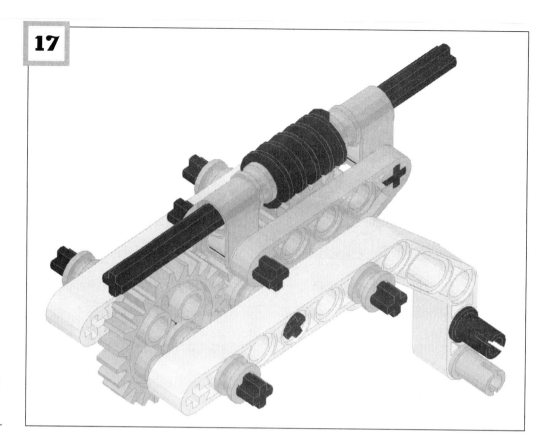

Secure the cross axles with half bushings and attach a black friction pin and a gray axle pin to each bent liftarm as shown.

18

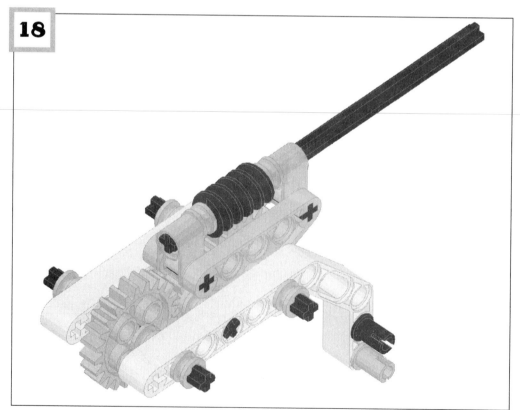

This completes one portion of the body. Repeat Steps 13 through 17 to build a second, identical unit, and attach the remaining gear assembly to it, as shown.

19

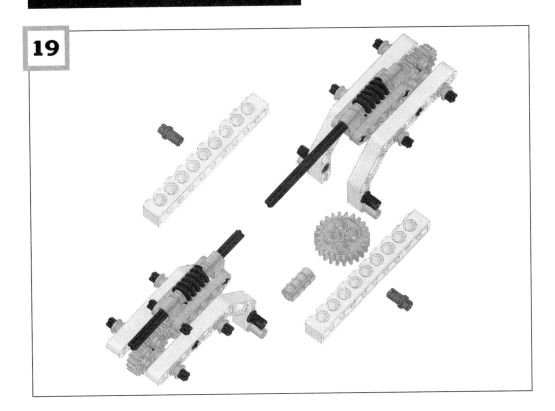

Gather two 1x10 beams and the two gear assemblies as shown.

20

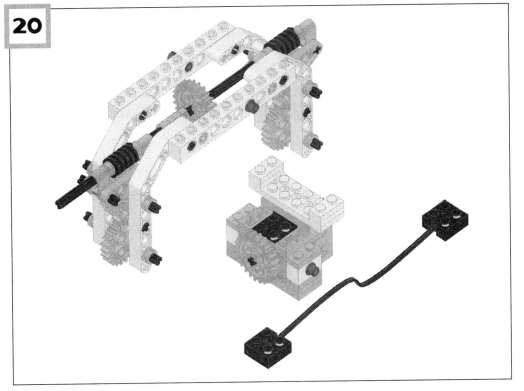

Attach a 24-tooth gear to the L12 worm gear assembly's cross axle. Make sure you have aligned the cross axles that pass through the assemblies' 24-tooth gears (see the Note after Step 7 in Model 3), then connect the assemblies with an axle joiner. Adjust the cross axles so that, when connected, they fit the 1x10 beams, then place the TECHNIC pins in the 1x10 beams as shown.

21

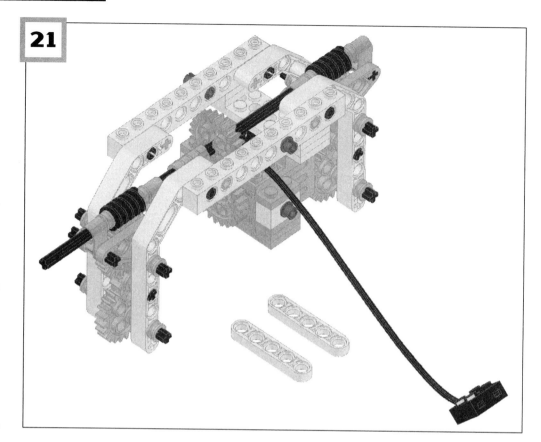

Attach a connecting lead to the motor, then place the motor beneath the body. Be sure to adjust the body's 24-tooth gear to mesh with the motor's 24-tooth gear, or the robot's legs will not get enough power and you may ruin the gears!

22

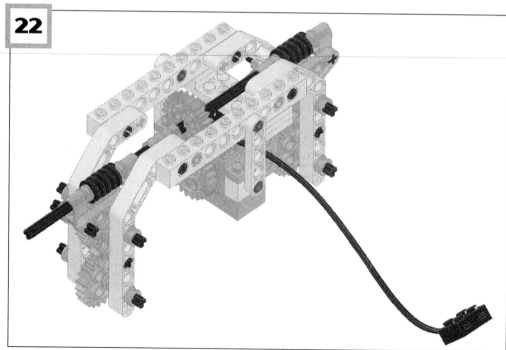

Attach 1x5 liftarms to the TECHNIC pins extending from the motor and the 1x11.5 bent liftarms to secure the body and motor.

23

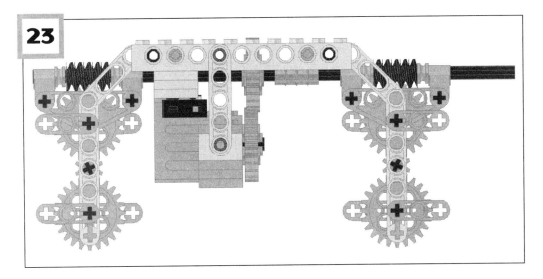

Position the 1x3 liftarms so that the left front and right rear legs have the same orientation, and the remaining two legs are offset by 180 degrees, as shown.

24

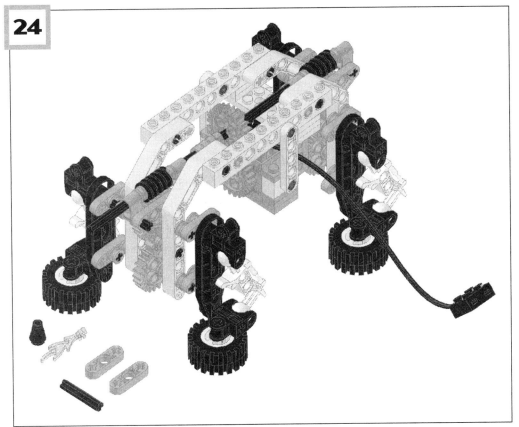

Attach the four legs by inserting their protruding cross axles through the middle holes of the 1x3 liftarms and securing the axles with half bushings.

25

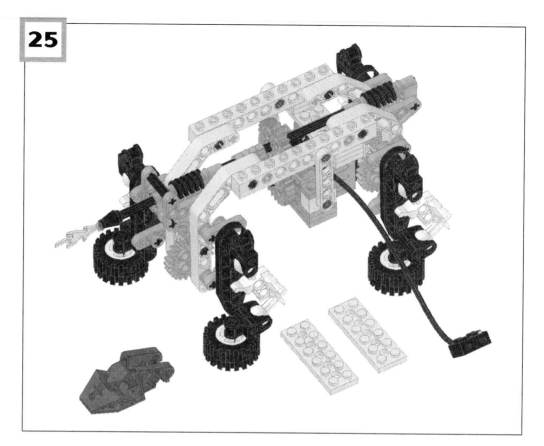

Attach 1x3 liftarms to the front of the body as shown. Attach the flame.

26

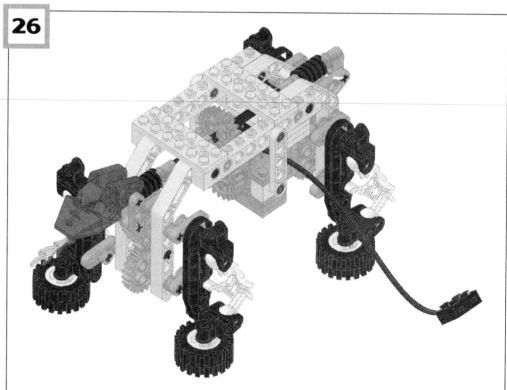

Attach the LEGO TECH-NIC SLIZER head to the 1x3 liftarms added in Step 25 and two 2x6 plates for mounting the RCX as shown.

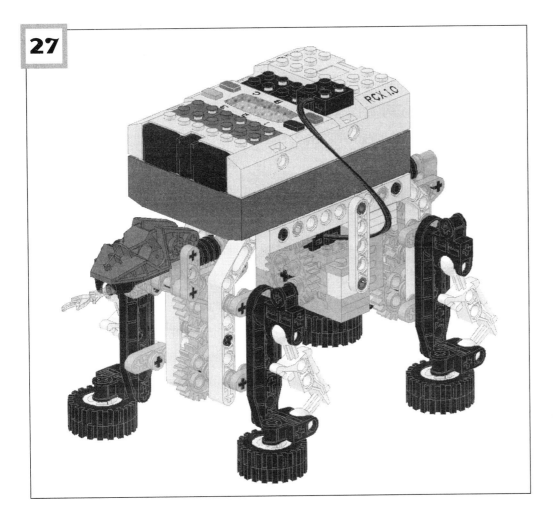

Finally, mount the RCX and connect the lead extending from the motor to output A of the RCX to complete the LEGOsaurus.

You can vary the look of your LEGOsaurus in many ways. One simple addition would be to add a tail by inserting a cross axle into a flexible tube and attaching it.

PROGRAM

This robot will move according to the default program in the RCX. Use the "Prog" button to select 1 and press the "Run" button to start the program.

TIP

Question: *Can we build a one-legged robot with only one gear and one cross axle?*

Answer: *Yes, we can build a top!*

Good tops, as we know from experience, rotate for extended periods without wobbling, but only if their weight is balanced and their axis precisely centered. If you build a top like the one shown here, using only LEGO pieces, it should be a very good one. Since LEGO pieces are created to very precise specifications, they can be used to build an excellent top.

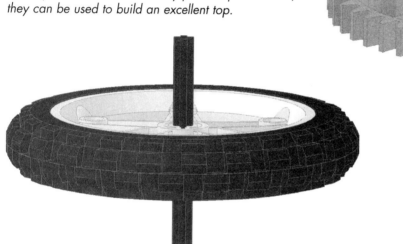

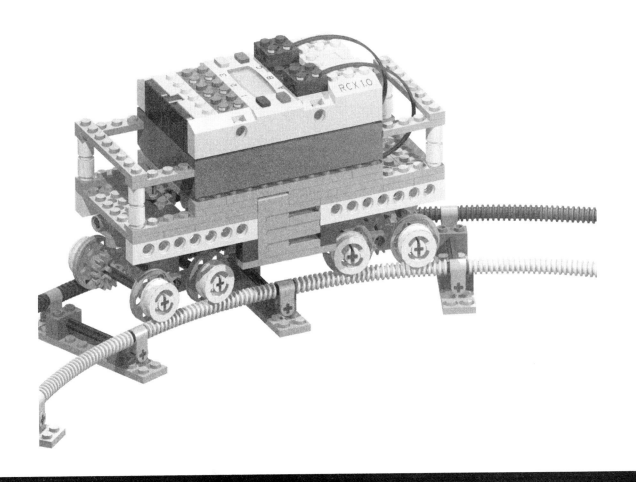

MODEL 5: TRAIN

In the world of LEGO MIND-STORMS robots, we can build almost anything if we try hard enough. In this chapter, we'll build a "train" and rails for it to run on. Although LEGO has produced a pre-built set of LEGO bricks for a train that runs on rails, it is, unfortunately, hard to find. Therefore, let's see how realistic a train we can build by combining ordinary LEGO bricks and parts.

The main challenge we face with this project is how to build the rails and wheels. One solution is to use flexible tubing for the rails. Because this tubing bends freely, we can easily build curved rails as well as straight ones. Also, although we could use wheel hubs for the wheels, let's place pulleys on the inner sides of the wheels to keep them from derailing while the train is running; this construction will produce a wheel shape like that of an actual train wheel.

PARTS LIST

TYPE	SIZE	QUANTITY
Train car		
Brick	1x1	4
Brick	1x1 round	8
Beam	1x8	4
Plate	1x4	6
Plate	1x6	4
Plate	1x8	6
Plate	1x10	2
Plate	2x4	5
Plate	2x8	6
Bracket	2x2–2x2	2
Liftarm	1x5	5
TECHNIC triangle		2

PARTS LIST (CONT.)

TYPE	SIZE	QUANTITY
Cross axle	2	4
Cross axle	5	4
Cross axle	6	2
Cross axle	8	4
Gear	12-tooth bevel gear	4
Gear	16-tooth	6
Gear	24-tooth	2
Gear	Worm gear	2
Pulley	Medium	2
Axle joiner		2
Perpendicular axle joiner	1x2	4
Half bushing		6
Bushing		2
Motor		2
Connecting lead	Short	2
Wheel hub	Small	8
CROSSTIES		
Brick	1x2 with axle hole	2
Plate	2x10	1
Cross axle	8	1
Perpendicular axle joiner	1x2	2
Friction pin	3	2
RAILS		
Flexible tube		2

BUILD THE WHEEL ASSEMBLY

1

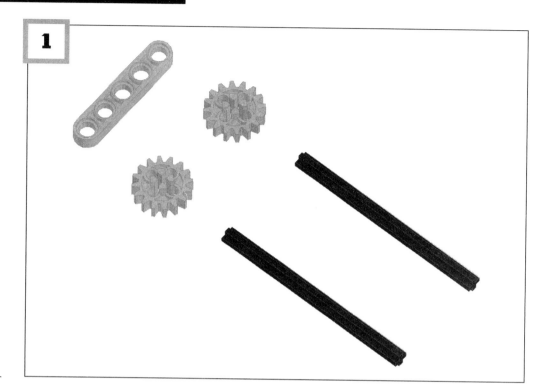

Gather the parts for the wheel assembly.

2

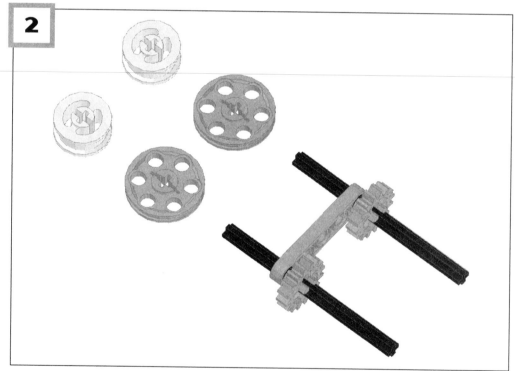

Attach two 16-tooth gears to two L8 cross axles and pass the cross axles through the 1x5 liftarm.

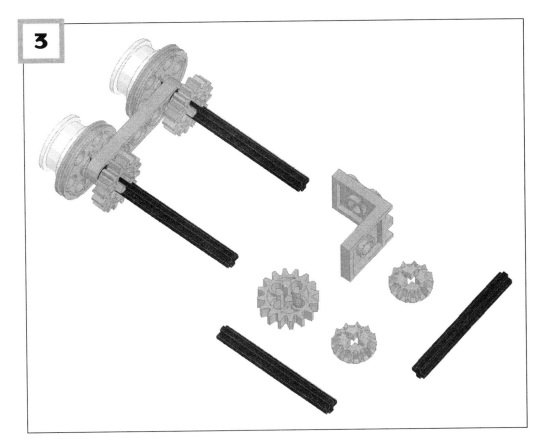

Attach a pulley and wheel hub next to the liftarm, on the outer side of each cross.

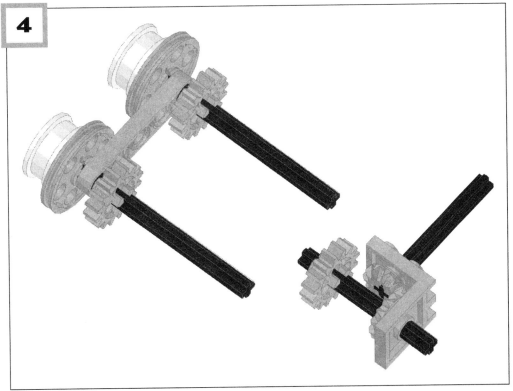

Slide a 16-tooth gear and a 12-tooth bevel gear onto an L5 cross axle and pass the cross axle through one side of a 2x2–2x2 bracket. Pass another L5 cross axle through the top of the 2x2–2x2 bracket, attach a 12-tooth bevel gear, and align it so that it meshes at a right angle with the first bevel gear. (You'll need to pass the second cross axle through the bracket before you can put the 12-tooth bevel gear on the axle.)

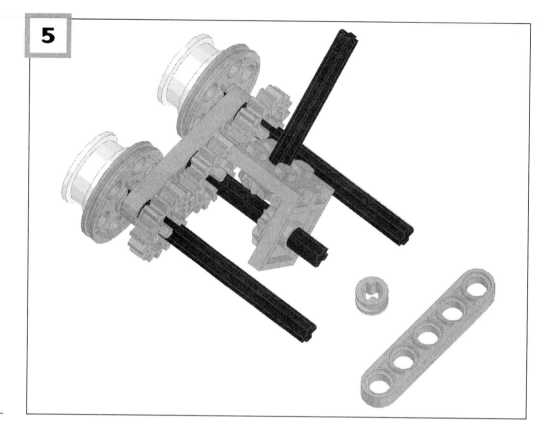

Attach the unit you created in Step 4 to the center of the liftarm so that the three 16-tooth gears mesh properly.

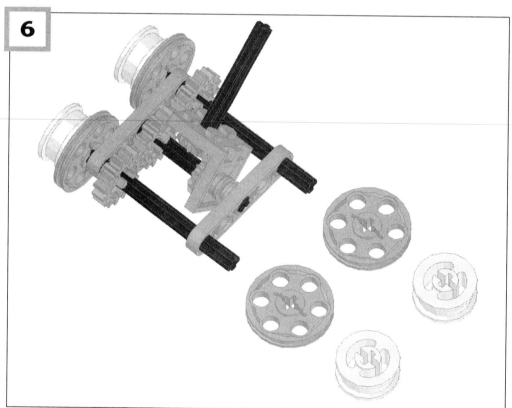

Attach a half bushing to the center L5 cross axle, then slide another 1x5 liftarm onto the cross axles.

7

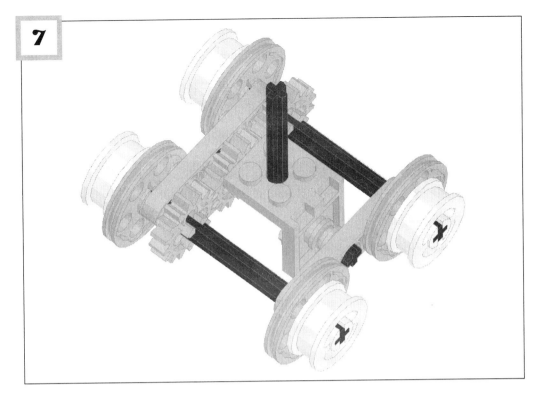

Attach pulleys and wheel hubs as described in Step 2. Straighten the bracket so its axle is perpendicular to the ground. This completes the wheel assembly.

BUILD THE WORM GEAR ASSEMBLY

8

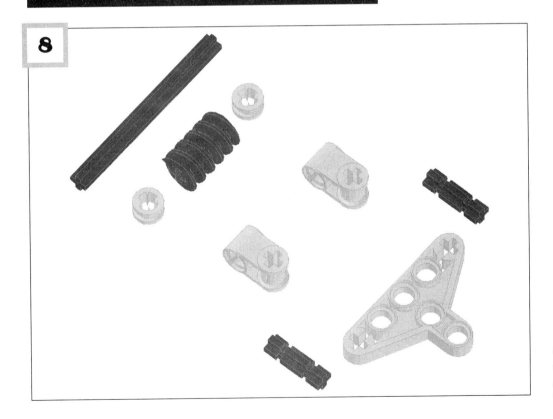

Gather the parts for the worm gear assembly as shown.

9

Insert two L2 cross axles into the TECHNIC triangle and attach perpendicular axle joiners to them as shown. Pass an L6 cross axle through one axle joiner, then slide a half bushing, worm gear, and half bushing onto the cross axle, in the same order as written here. Finally, pass the L2 cross axle through the other perpendicular axle joiner.

BUILD THE FLATCAR

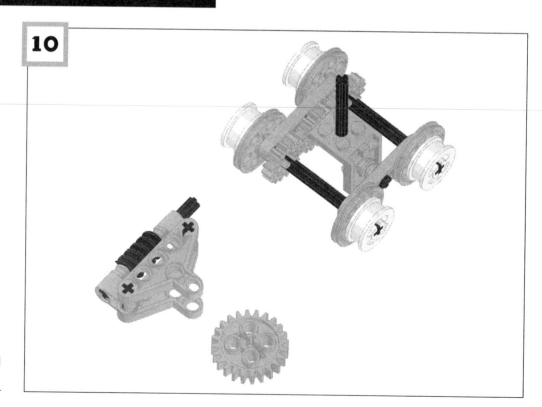

Collect the two assemblies created above and a 24-tooth gear.

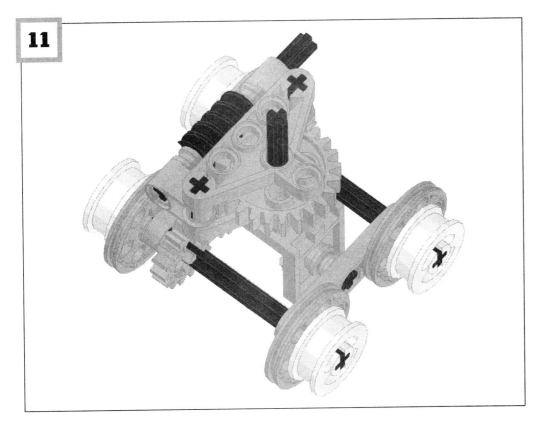

Connect the wheel and worm gear assemblies by inserting the 24-tooth gear between the two TECHNIC triangles; secure it by passing the wheel assembly cross axle through the triangles' center holes as shown.

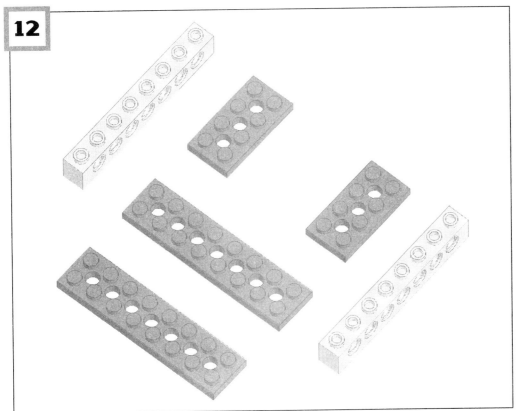

12

Gather two 1x8 beams, two 2x8 plates, and two 2x4 plates.

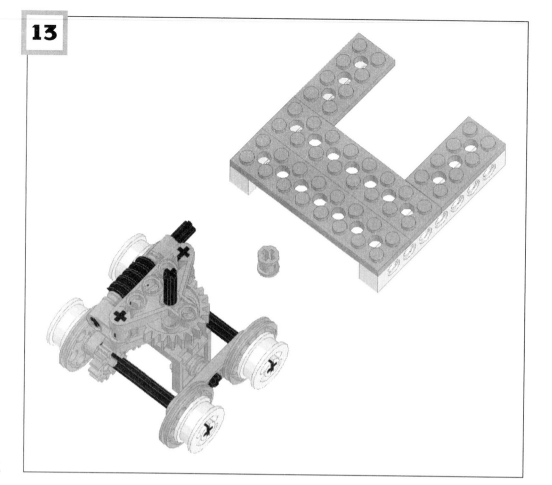

13

Attach the two 2x8 plates across the tops of the 1x8 beams and add the 2x4 plates as shown.

14

Pass the vertical cross axle of the flatcar through the middle hole of the second 2x8 plate and secure it from the top with a bushing. Make sure that the indented side of the bushing is at the top and the circular side is at the bottom. (If the indented side is at the bottom, the bushing will mesh with the plate and the wheels will not turn.) Also be sure that the cross axle that passes through the worm gear is parallel to the 1x8 beam.

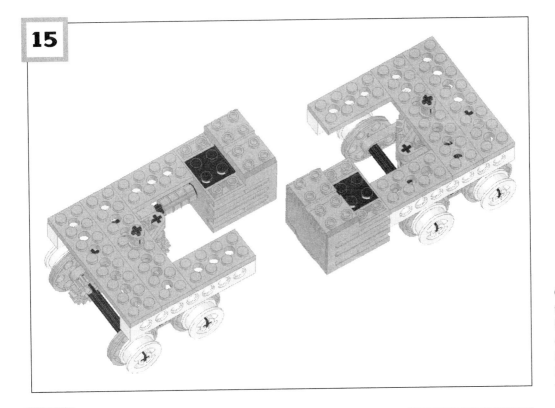

Connect the motor to the worm gear assembly's cross axle with an axle joiner, then follow Steps 1 through 14 to build an identical unit.

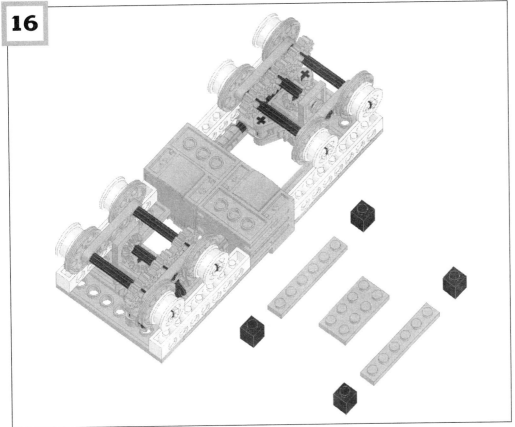

Join the two units so that the motors sit side by side, then turn the entire assembly over (upside down) so that you'll be ready to secure the frame bottoms.

17

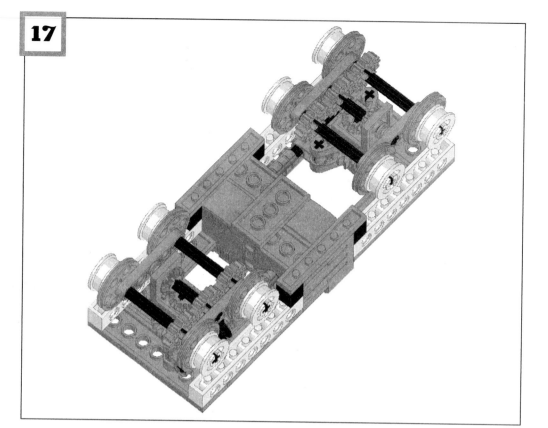

Connect the two motors with a 2x4 plate and attach four 1x1 bricks to the frame, then secure the motors with two 1x6 plates.

18

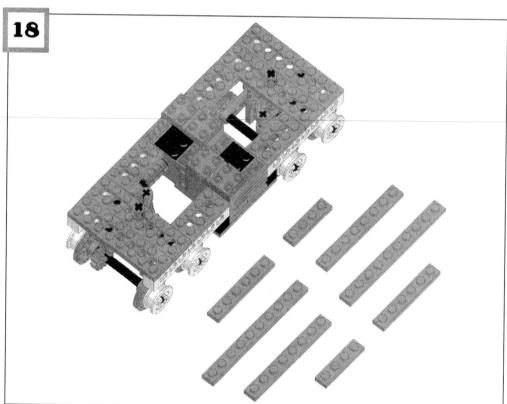

Turn the frame back upright to continue working on the top, then gather the plates for the next step.

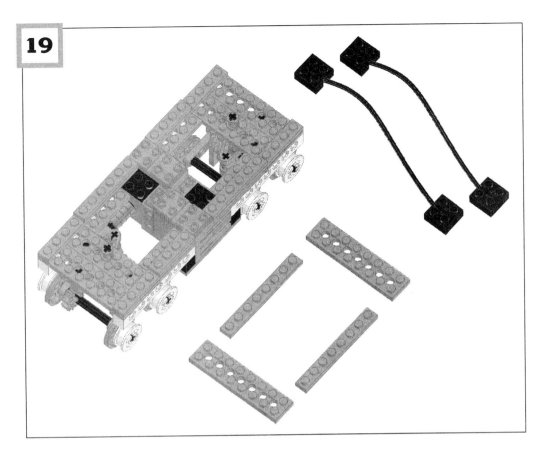

19

Attach two 1x10 and
two 1x8 plates along
the edges of the frame,
then attach 1x6 plates
to the 1x10 plates and
1x4 plates on top of
the 1x8 plates.

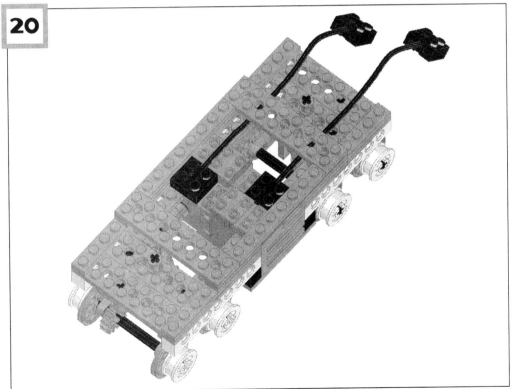

After attaching both
connecting leads to the
motors, use 2x8 and
1x8 plates to reinforce
the top of the assembly.

21

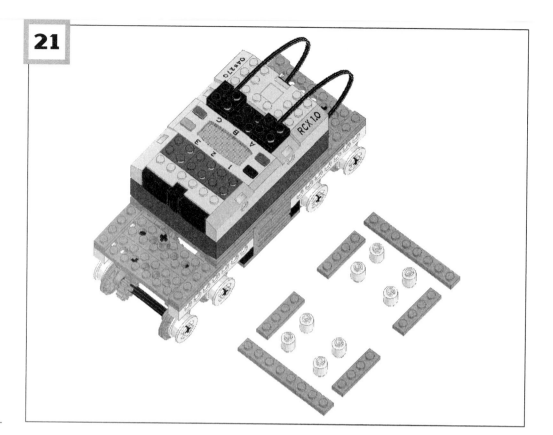

Mount the RCX in the center of the train and attach the connecting leads to outputs A and C.

22

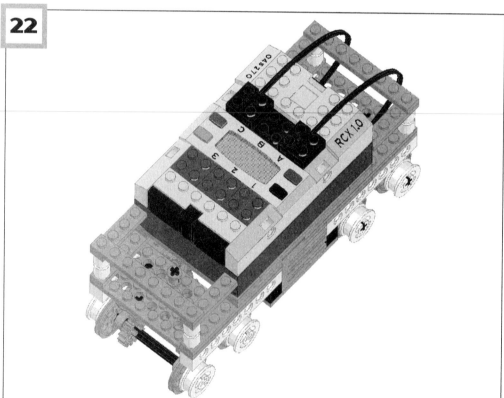

Build four sets of two stacked 1x1 circular bricks and combine them with 1x8 and 1x4 plates to build railings. This completes the train.

23

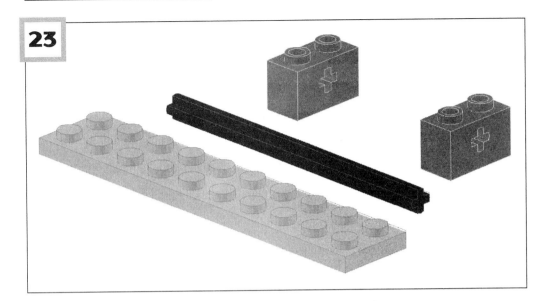

Gather the parts for the rails as shown.

24

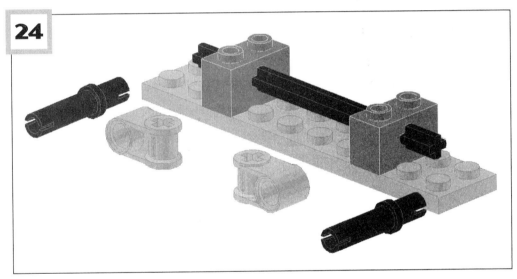

Pass an L8 cross axle through two 1x2 bricks with axle holes and attach them to the top of the 2x10 plate.

25

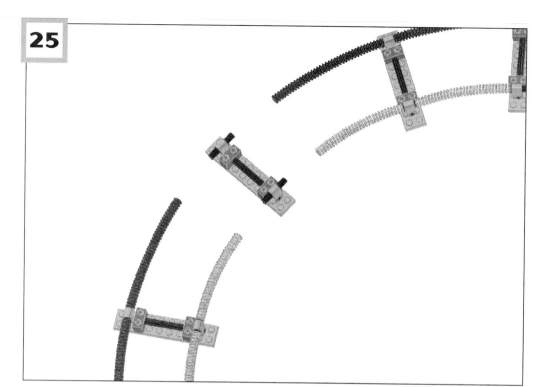

Attach perpendicular axle joiners to both ends of the cross axle, then pass friction pins of length 3 through them. This completes one crosstie.

26

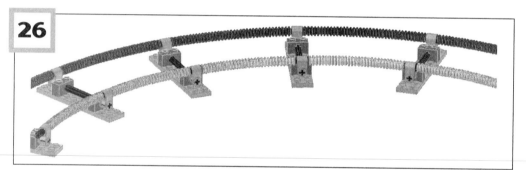

Build the rails by sliding flexible tubes onto the friction pins extending from the crossties. If both flexible tubes are the same length, the rails will be straight; if they are different lengths, the rails will curve.

27

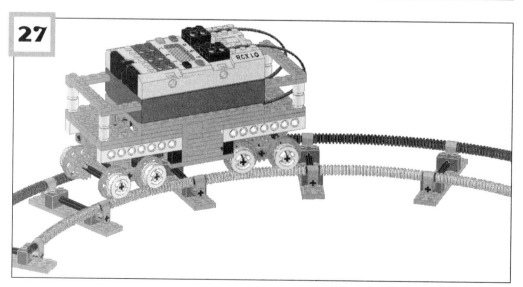

Place the train on top of the rails.

PROGRAM

To move the train, create a program with the RCX code that turns on outputs A and C.

Alternatively, if you stack the two leads from the motors and connect them to output A, you can move the train by using the RCX's default program. Use the "Prog" button to select Program 1 and then press the "Run" button to start the default program.

GOING FURTHER

Use a light sensor to measure the train's position by counting crossties. Try building a station and programming the train to advance, stop, and back up automatically.

MODEL 6: WALKER

Whenever anyone starts using LEGO MINDSTORMS kits, a two-legged walking robot is one of the first things they want to build. Although walking on two legs is extremely simple for humans, it's quite difficult to accomplish with a machine. Let's give it a try, though, and build a two-legged Walker out of LEGO MINDSTORMS parts.

Why is it difficult to walk on two legs? When walking, a two-legged robot is almost always balancing on one leg, so it will topple over if its center of gravity is not located over the grounded leg. We humans remain upright by unconsciously shifting our body weight with each step, thus shifting our center of gravity.

Figure 6-1 should help you to visualize the problem. The gray and white bars represent the Walker's feet, and the black dots represent its center of gravity when viewed from above. The robot begins at (a), lifts its right foot (b), moves its right foot forward (c), and finally lowers its right foot to the ground (d). When both feet are touching the ground, at (a) or (d), the Walker's center of gravity is supported by its two feet. However, when the Walker is standing on one foot only, as in position (b) or (c), it falls over, because its center of gravity is not located over the grounded foot.

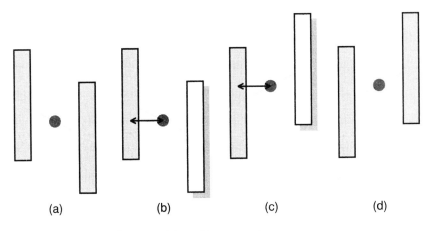

Figure 6-1: Center of gravity while walking on two legs

To solve this problem, we need to balance the Walker by positioning its center of gravity over its grounded foot. To do so, we'll design a specialized foot, as shown in Figure 6-2.

By making C-shaped feet, we support the center of gravity, even when the Walker is standing on only one leg. For example, see stages (b) and (c), showing a raised right foot. With the right foot raised as shown in these stages, our C-shaped left foot would support the center of gravity (indicated by the black dots) as long as it is within the outer edges of the foot and inside the dotted line. Because the Walker's center of gravity is always within the range enclosed by the foot and this line, it can be supported by only one foot.

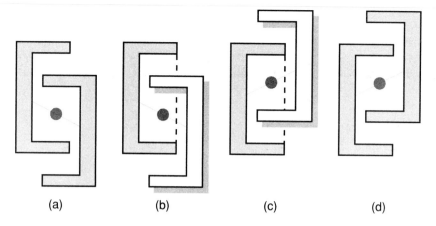

(a)   (b)   (c)   (d)

Figure 6-2: *Center of gravity with improved feet*

The sequence of photographs below (Figure 6-3) shows how the Walker's movement applies this mechanism.

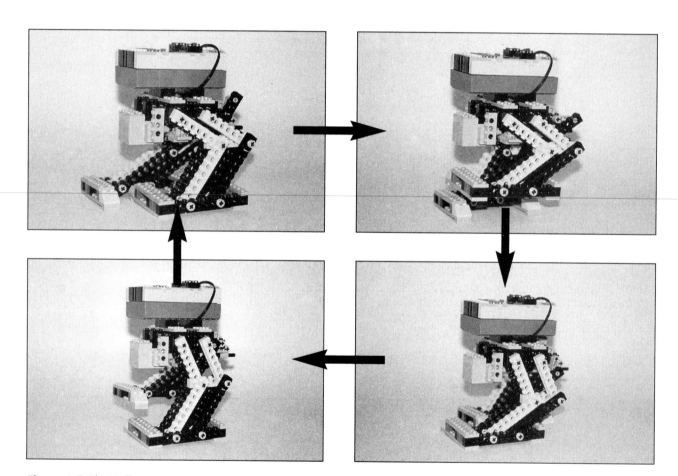

Figure 6-3: *The Walker in action*

PARTS LIST

TYPE	SIZE	QUANTITY
Brick	2x4	2
Beam	1x2	2
Beam	1x4	2
Beam	1x6	6
Beam	1x8	6
Beam	1x10	8
Beam	1x12	6
Beam	1x16	2
Brick	1x2 with axle hole	8
Brick	2x2 slanted cut	4
Plate	1x2	4
Plate	1x2 with rail	4
Plate	1x8	2
Plate	2x8	2
Plate	2x10	6
Liftarm	1x3	4
TECHNIC triangle		2
Gear	24-tooth	1
Worm gear		1
Cross axle	2	2
Cross axle	5	2
Cross axle	6	11
Cross axle	8	4
Cross axle	12	1
Axle joiner		1
Perpendicular axle joiner	1x2	2
Friction pin	2	2
Friction pin	3	2
TECHNIC pin	1/2	8
TECHNIC axle pin		2
Half bushing		26

PARTS LIST (CONT.)

TYPE	SIZE	QUANTITY
Bushing		10
Motor		1
Connecting lead	Short	2
Light sensor		1
Fiber optic light unit		1
Fiber		8

ASSEMBLY PROCEDURE

BUILD THE LEGS

NOTE *If you don't have enough long beams for Steps 1 through 9, follow the alternative method directly after Step 9 below.*

1

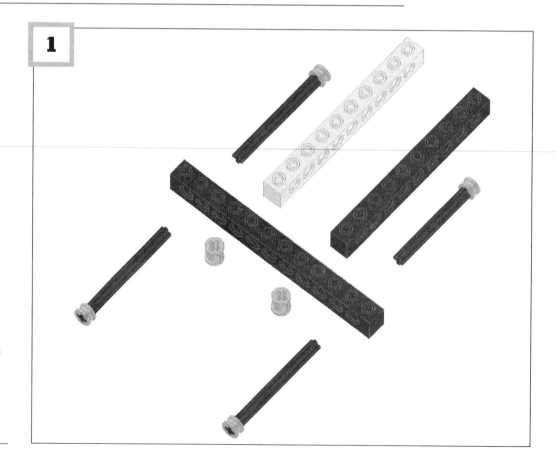

Assemble the parts shown, including three L6 and one L5 cross axle (shown at upper right). Attach one half bushing to each cross axle.

2

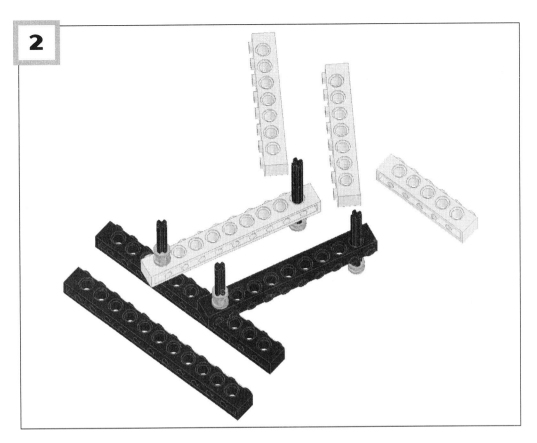

Attach the two 1x10 beams to the top of the 1x12 beam (black), using two of the L6 cross axles and the two bushings. Assemble the other L6 cross axles as shown.

3

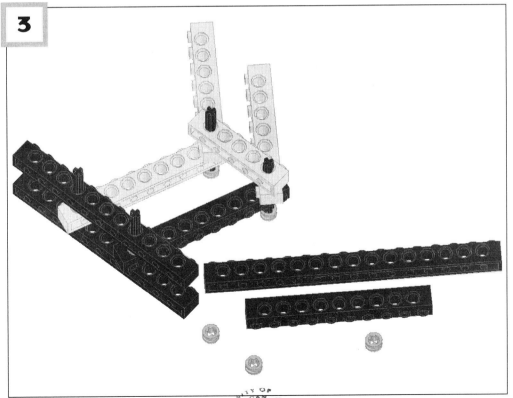

Slide two 1x8 beams to the L6 cross axles at the ends of the two 1x10 beams, then attach a 1x6 beam, sideways, to the top of the 1x8 beams as shown. Slide a 1x12 beam to the other pair of L6 cross axles, as shown in the foreground.

4

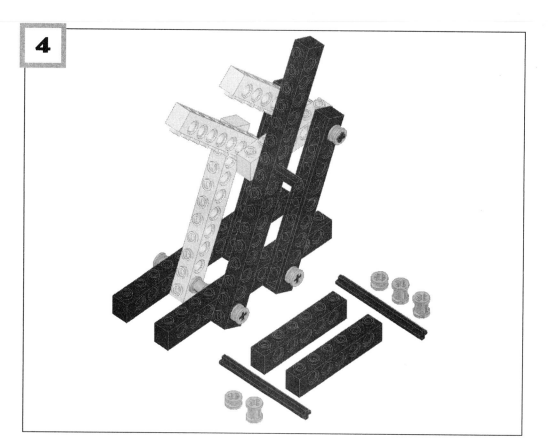

Attach a 1x16 beam and a 1x10 beam to the top of this assembly and secure them with half bushings as shown.

5

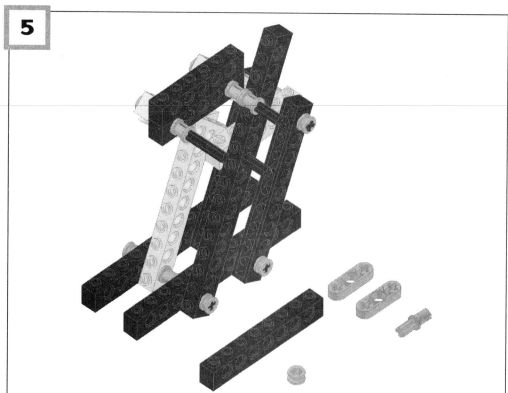

Use L6 cross axles to connect a 1x6 beam to the ends of the two 1x8 beams at the top of the leg, to form part of the Walker's body. Stack another 1x6 beam on top of this 1x6 beam, and attach bushings to the protruding cross axles as shown.

Model 1: *Rolling Car*

Model 2: *Centipede*

Model 3: *Water Skater*

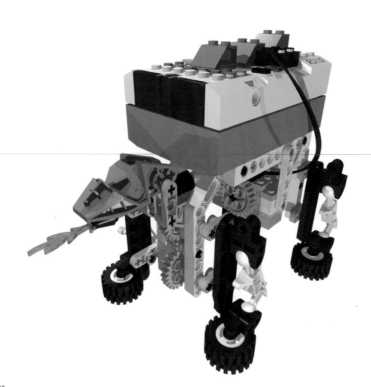

Model 4: *LEGOsaurus*

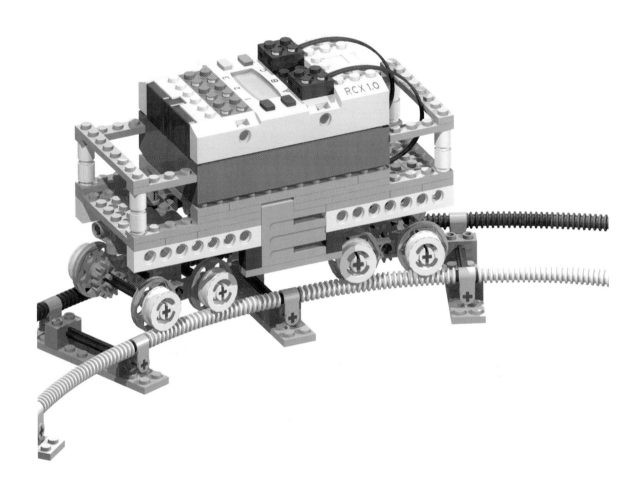

Model 5: *Train*

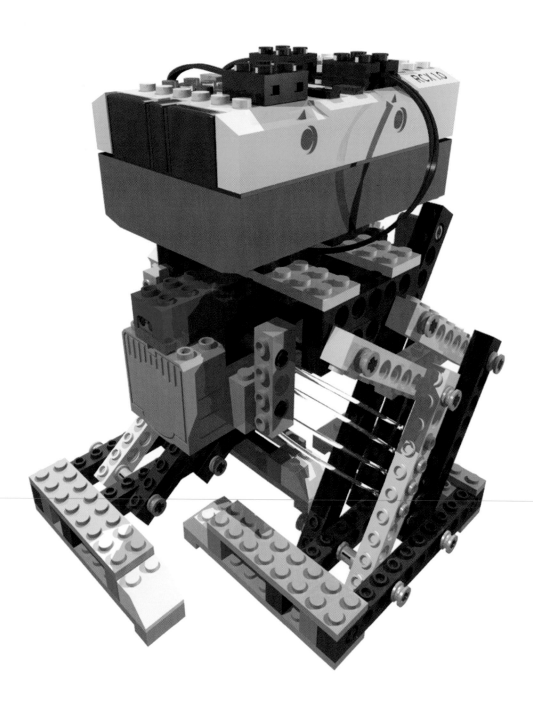

Model 6: Walker

Model 7: *Climber*

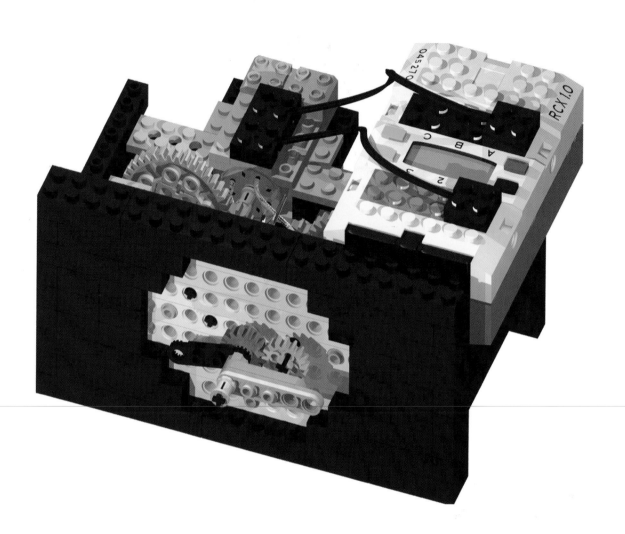

Model 8: LEGO Clock

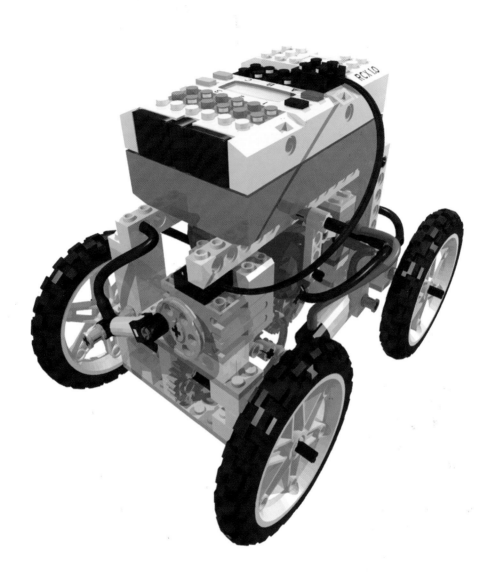

Model 9: Pneumatic Engine Car

Model 10: *Beetle*

6

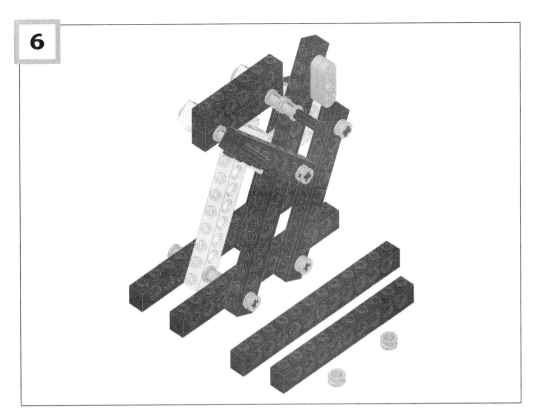

Slide a 1x8 beam to the top left cross axle and the cross axle protruding from the 1x16 beam, then use a TECHNIC axle pin to attach two 1x3 liftarms to the end of the 1x16 beam. (The motor will rotate these liftarms to move the leg.)

ALT

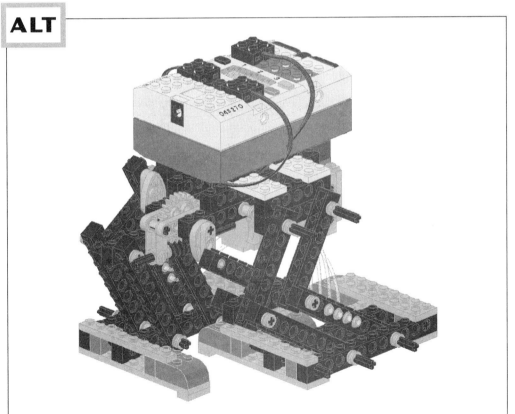

ALTERNATIVE: If you haven't enough liftarms to complete Step 6, you can use a TECH-NIC cam instead. The Walker shown here, from the rear, uses a stacked liftarm and cam for each leg.

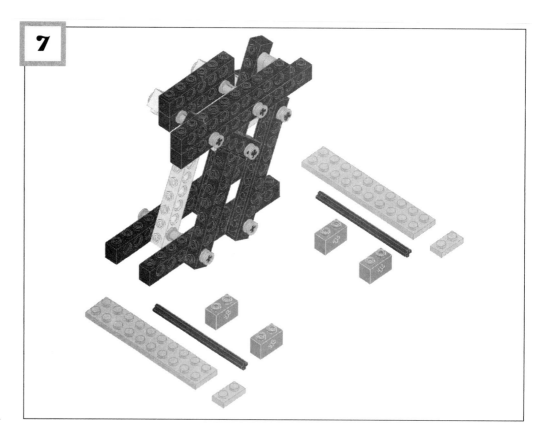

7

Insert the cross axles at the top of the leg through the third and seventh holes of a 1x12 beam, then add a 1x10 beam on top of the 1x12 beam. Secure the cross axles with half bushings to complete the basic leg structure.

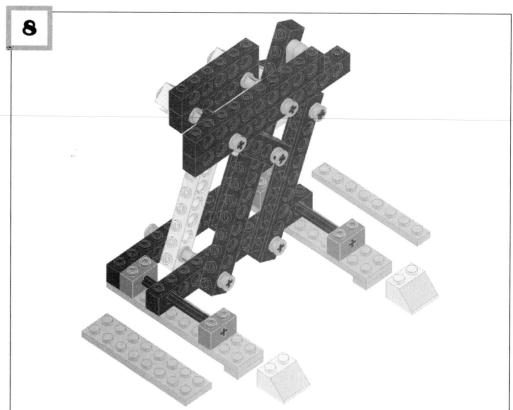

8

Let's build the feet. Arrange two 2x10 plates on either side of the leg, then pass two L8 cross axles through the plates and through the ends of the leg's bottom beams to act as "bones"—structural supports. Attach the cross axles to the plates with 1x2 bricks with axle holes.

9

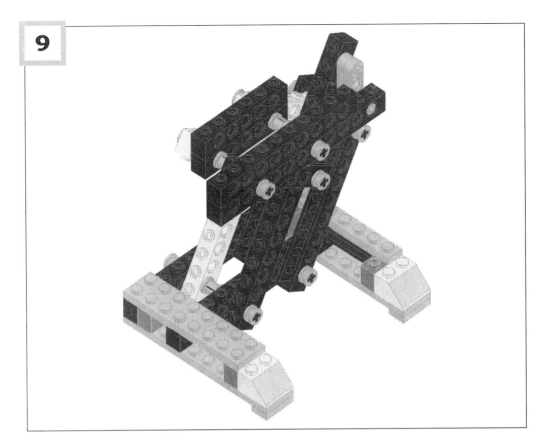

Attach a 2x8 plate to the top of the front "bone" (shown at left) and a 1x8 plate to the top of the back "bone" (shown at right). Decorate the tips of the toes with two yellow, slanted, cut bricks as shown.

ALT

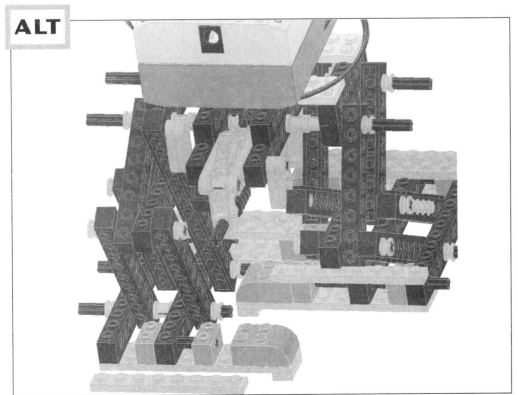

ALTERNATIVE: You can use two L4 cross axles, instead of L8 cross axles, for the bone portion of the foot, as shown here.

ALTERNATIVE TO STEPS 1-9: If you don't have enough long axles for Steps 1 through 9, you can adjust their lengths, using axles of the lengths indicated by the numbers shown outside the parentheses in the figure (the black dots represent the axles). If you are building the robot using only the LEGO MINDSTORMS Robotics Invention System set, use axles of the lengths indicated by the numbers inside the parentheses, and use half bushings on the outside of the leg. The axles will protrude a little on the outside, but the Walker will still function.

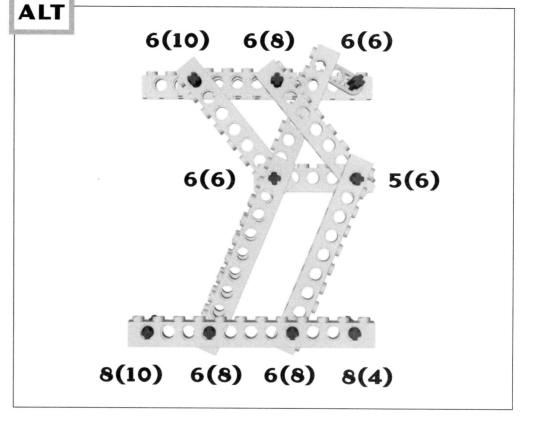

10

Build another leg, identical to the first but symmetrically reversed. Make sure to secure all of the axles and bushings so that the legs will not hit each other when they move forward.

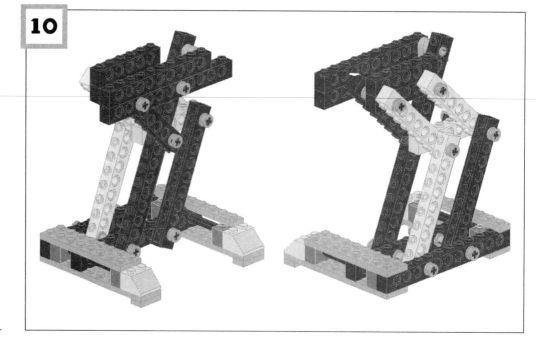

11

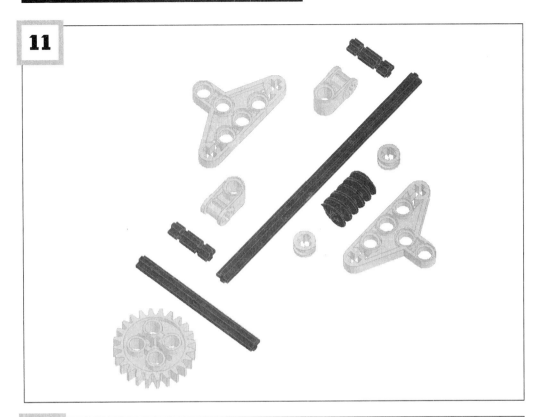

Assemble the pieces shown to build the gear assembly which will transmit power to the legs.

12

Use two L2 axles to attach two perpendicular axle joiners to the TECHNIC triangle. Attach a worm gear and half bushings by passing an L12 cross axle through the two axle joiners and the gear. Next, pass an L6 cross axle through both a 24-tooth gear and the TECHNIC triangle, meshing the gear's teeth with the worm gear.

13

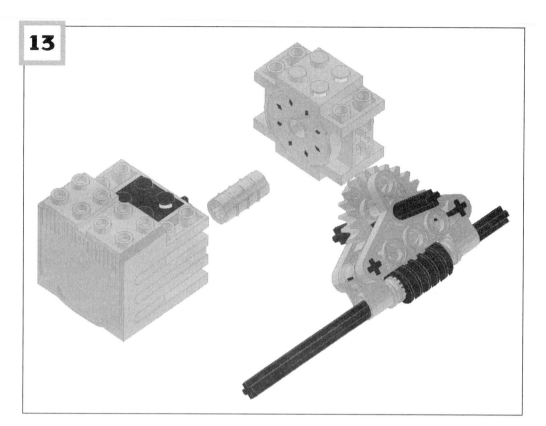

Attach another triangle to complete the gear assembly as shown.

14

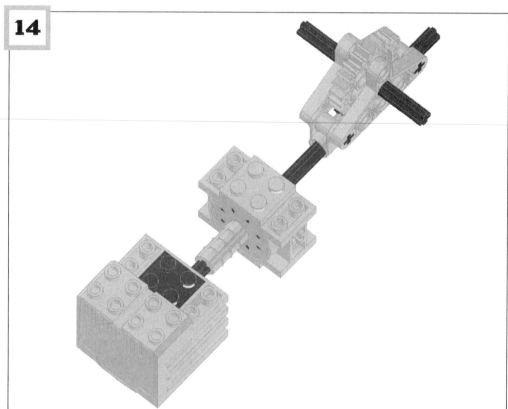

Use an axle joiner to connect the motor and gear assembly. If you have a fiber optic light unit, set it between the motor and gear assembly and pass the L12 cross axle through it. This completes the motor and gear assembly unit.

If you are building the Walker using only the LEGO MINDSTORMS Robotics Invention System and have too few TECHNIC triangles, you can use the following method to build the gear assembly.

1

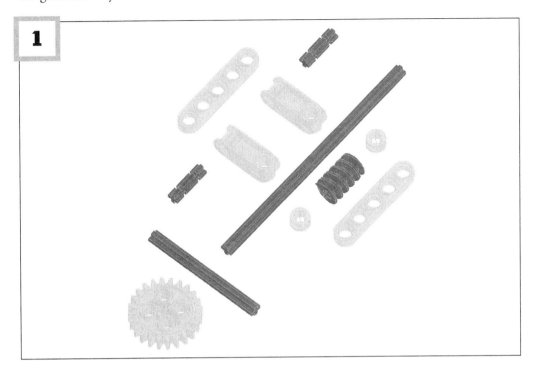

Gather the parts shown in the figure.

2

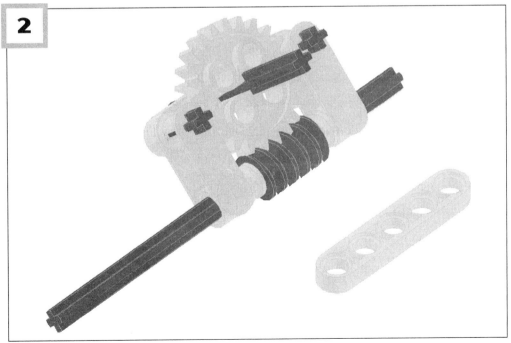

Use 1x5 liftarms and 1x3 perpendicular axle joiners to mesh the worm gear with the teeth of the 24-tooth gear.

3

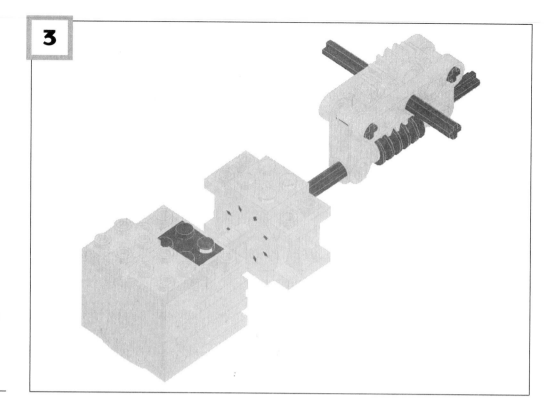

Connect the gear assembly, motor, and fiber optic light unit to complete this section of the robot. (If you do not have a fiber optic light unit, just connect the gear assembly and motor.)

CONNECT THE LEGS AND GEAR ASSEMBLY

15

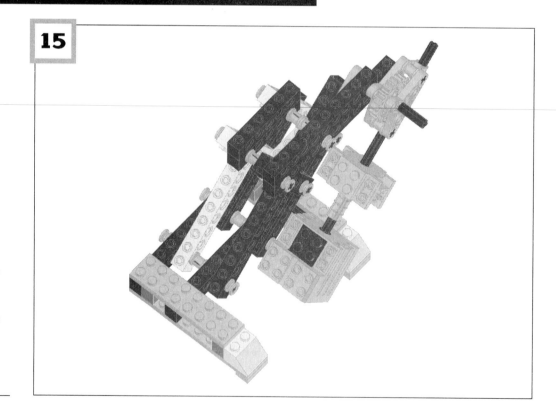

Pass the L6 cross axle protruding from the 24-tooth gear through the 1x12 beams, then connect it to the 1x3 liftarms. (Make sure the worm gear is on the bottom.)

16

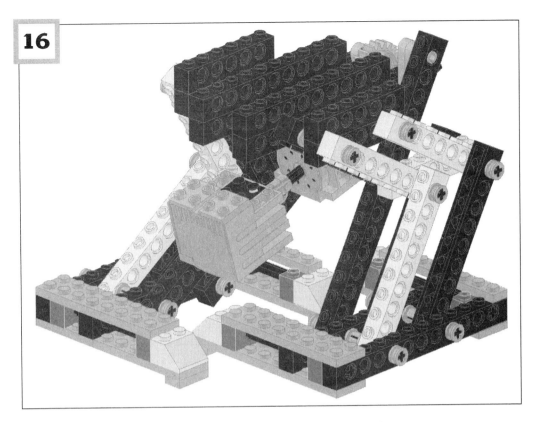

Repeat the same steps above for the other leg, offsetting the orientation of the liftarms on the left and right legs by 180°.

17

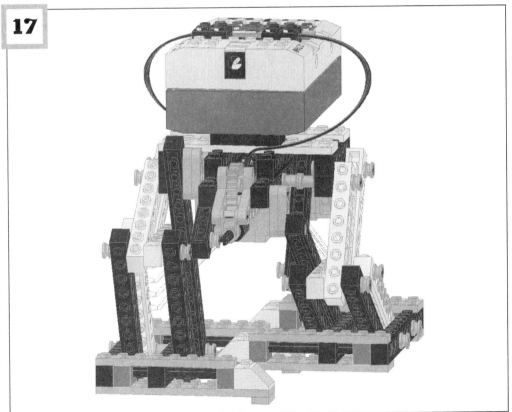

This figure shows the completed robot from the rear. Check that the liftarms on the left and right legs are positioned in opposite directions as shown.

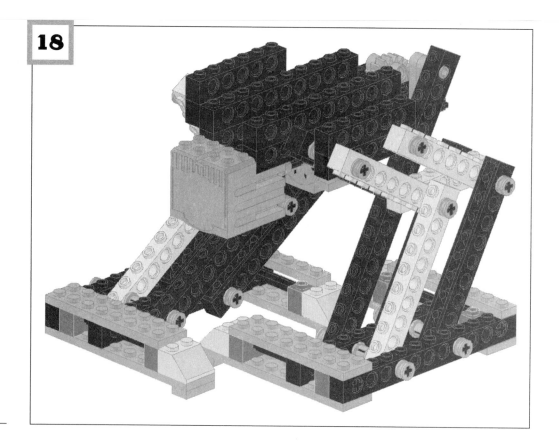

18

Carefully attach the
motor and fiber optic
light unit to the body.

REINFORCE THE MOTOR

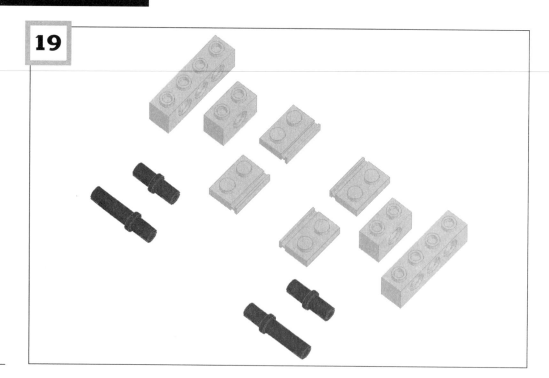

19

Collect the parts you'll
need to reinforce the
motor.

20

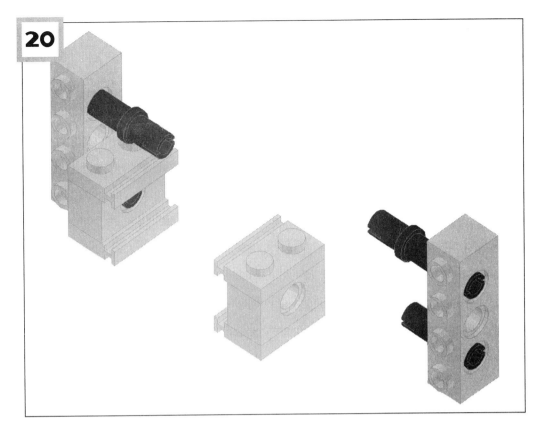

Build the parts to secure the motor on the left and right sides, as shown.

21

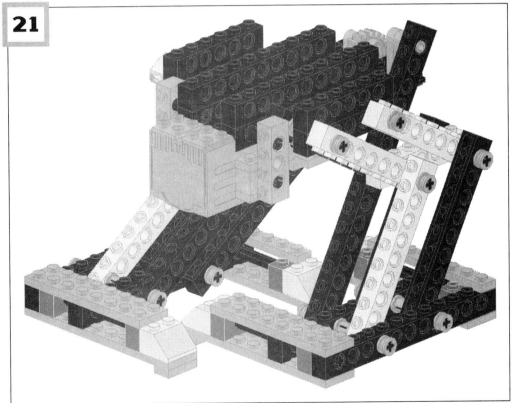

Use these parts to attach the motor to the beams on both sides of the body.

22

Connect the wiring leads to the motor and fiber optic light unit, then attach two 2x10 plates to reinforce the tops of the legs. Place a 2x4 brick on top of each 2x10 plate to build a base for mounting the RCX.

23

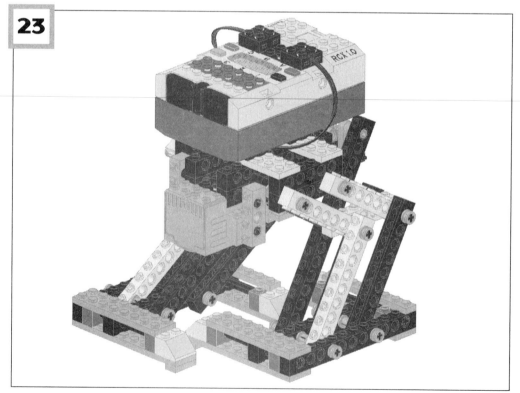

Mount the RCX on the 2x4 bricks. Connect the lead from the motor to output A of the RCX and the lead from the fiber optic light unit to output C of the RCX. This completes the basic structure.

24

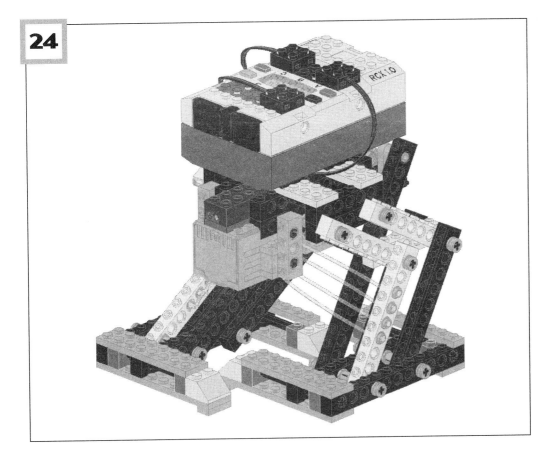

Try attaching fibers to the legs as decorations. If you attach a light sensor on top of the motor in the front as shown here, you can program the robot in various ways.

PROGRAM

The Walker does not require a complicated program. Following are two types: One is a simple program to make the robot walk; the other uses the light sensor.

WALKING

A simple program that turns on outputs A and C of the RCX will make the robot walk slowly, one step at a time, and emit a red light through the decorative fibers (if you used the fiber optic decorations and light unit). Because this program is set by default in the RCX, you can run it by simply using the "Prog" button to select Program 1.

SENSING THE BRIGHTNESS OF THE SURROUNDINGS

The program in Figure 6-4 reads the light sensor value and causes the fibers to light up only when the surroundings are dark. The light sensor threshold value differs according to the brightness of the room, so test various values to determine when to make the fibers light up.

Changing the program so that the light sensor turns output A on or off will make the robot walk or stop depending on the brightness of its surroundings.

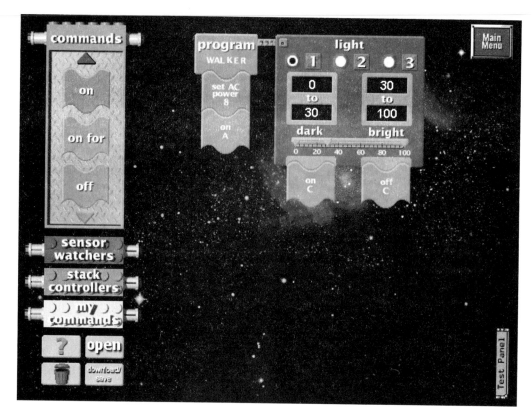

Figure 6-4: Program that uses the light sensor

TIP

RUBBER ERASER

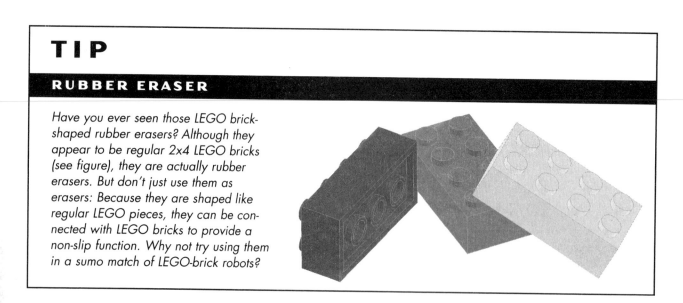

Have you ever seen those LEGO brick-shaped rubber erasers? Although they appear to be regular 2x4 LEGO bricks (see figure), they are actually rubber erasers. But don't just use them as erasers: Because they are shaped like regular LEGO pieces, they can be connected with LEGO bricks to provide a non-slip function. Why not try using them in a sumo match of LEGO-brick robots?

MODEL 7: **CLIMBER**

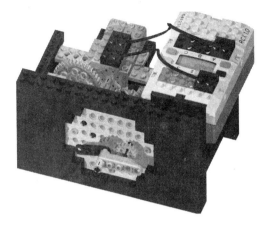

MODEL 8: **LEGO CLOCK**

MODEL 9: **PNEUMATIC ENGINE CAR**

MODEL 10: **BEETLE**

PART 3
ADVANCED

MODEL 7: CLIMBER

In this chapter we'll build an autonomous robot, the Climber, that senses and climbs over obstacles. Robots use various methods to climb over obstacles; ours changes its shape to do so.

We'll write an NQC program that makes the Climber's frame deform when it bumps into obstacles.

The Climber has eight wheels. The front, middle, and rear portions of its body fold by pivoting at the central set of four wheels, thus enabling it to climb over obstacles.

Figure 7-1 illustrates how this deformation works. First, when the robot bumps into an obstacle (a), it senses the obstacle using a touch sensor and deforms its frame to place its front wheels on top of the obstacle (b). It next folds its frame in the opposite direction (c), moves forward (d), then folds its frame back again (e) and continues onward (f). At this point, it returns its frame to horizontal, (g) and (h). To climb down, it uses its light sensor to detect that the ground is now farther away (that is, the front of the Climber is hanging over the edge of the obstacle) and repeats the previous steps in reverse—that is, (g) to (a).

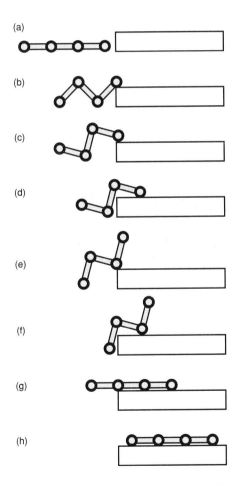

Figure 7-1: How the Climber deforms itself

The following sequence of photographs show the Climber in action (Figure 7-2), as it climbs over an obstacle.

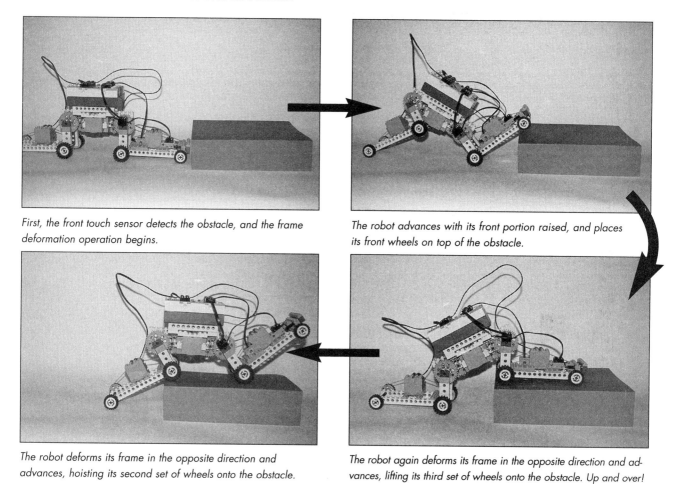

First, the front touch sensor detects the obstacle, and the frame deformation operation begins.

The robot advances with its front portion raised, and places its front wheels on top of the obstacle.

The robot deforms its frame in the opposite direction and advances, hoisting its second set of wheels onto the obstacle.

The robot again deforms its frame in the opposite direction and advances, lifting its third set of wheels onto the obstacle. Up and over!

Figure 7-2: The Climber in action

PARTS LIST

TYPE	SIZE	QUANTITY
Beam	1x2	21
Beam	1x6	3
Beam	1x8	2
Beam	1x10	2
Beam	1x16	5
Brick	1x2 with axle hole	7
Plate	1x2	4
Plate	1x2 with rail	4

PARTS LIST (CONT.)

TYPE	SIZE	QUANTITY
Plate	1x6	5
Plate	2x2	2
Plate	2x6	6
Plate	2x4	5
Bracket	2x2–2x2	4
Liftarm	1x7	4
Gear	8-tooth	2
Gear	24-tooth	1
Gear	40-tooth	4
Pulley	Medium	4
Cross axle	3	2
Cross axle	4	3
Cross axle	8	2
Cross axle	10	4
Half bushing		5
Bushing		8
Motor		3
Connecting lead	Short	3
Connecting lead	Medium	2
Light sensor		1
Touch sensor		1
Wheel hub		8
Rubber tire	Small	4
Rubber tire	Medium	4
Rubber band		4
Bushing		10
Motor		1
Connecting lead	Short	2
Light sensor		1
Fiber optic light unit		1
Fiber		8

ASSEMBLY PROCEDURE

1

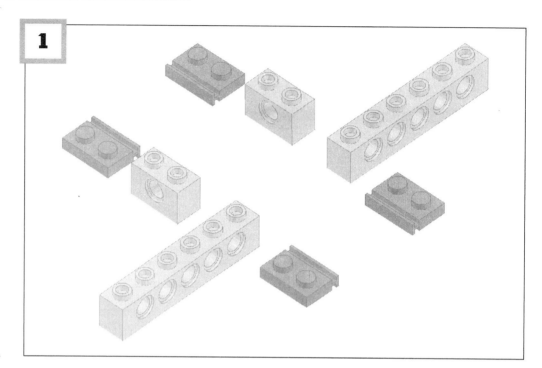

Gather the parts for the middle portion of the frame, as shown.

2

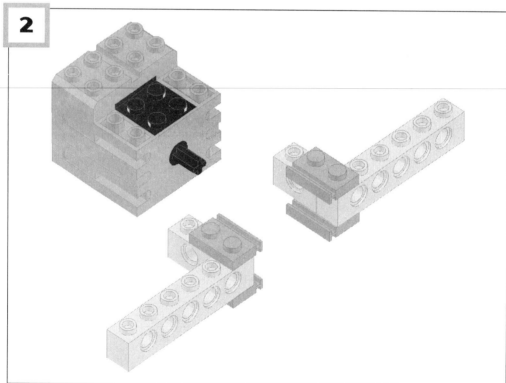

Attach 1x2 plates with rails to 1x6 and 1x2 beams as shown.

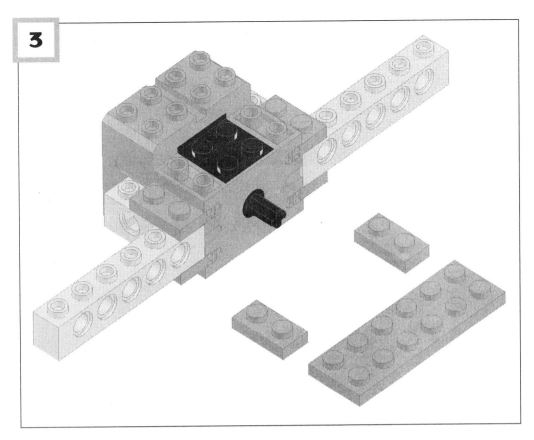

Slide the rail portions
into both sides of the
motor.

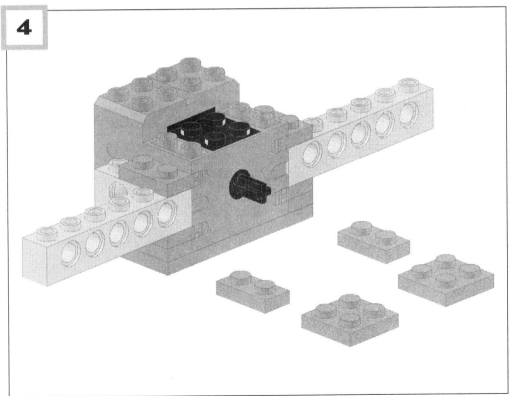

Place another set of 1x2
plates directly under the
bottom set of 1x2 plates
with rails and secure
them with a 2x6 plate.

Place another set of 1x2 plates directly over the top set of 1x2 plates with rails and secure them to the motor with 2x2 plates.

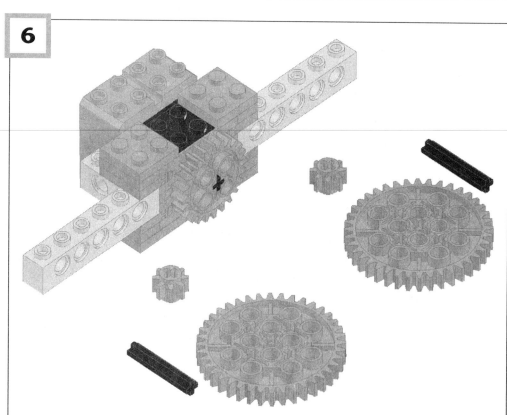

Slide a 24-tooth gear onto the motor's cross axle.

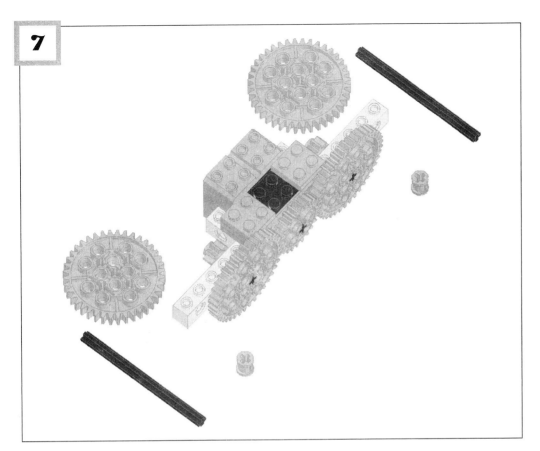

Slide two 40-tooth gears onto two L3 cross axles, then pass one cross axle through each of the 1x6 beams so that the 40-tooth gears and the 24-tooth gear mesh. Slide an 8-tooth gear onto each cross axle on the other side of the beam.

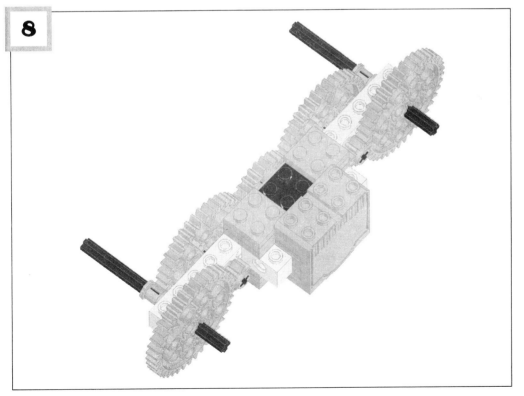

8

Slide two 40-tooth gears onto two L8 cross axles. Now, pass one cross axle through each of the 1x6 beams from the side opposite the first two 40-tooth gears and secure them with bushings. (This figure shows the unit from the opposite side.)

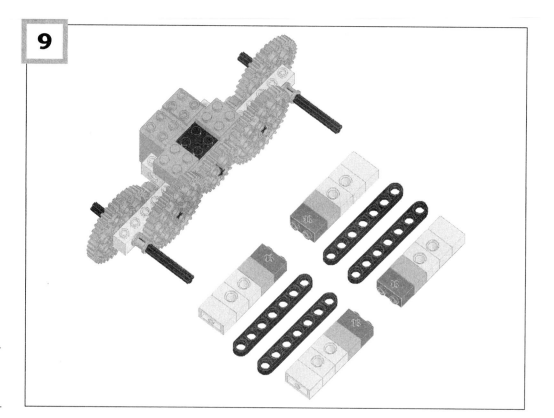

9

Build four pillars of four 1x2 bricks (with or without holes). Place a 1x2 brick with axle hole on top of each pillar.

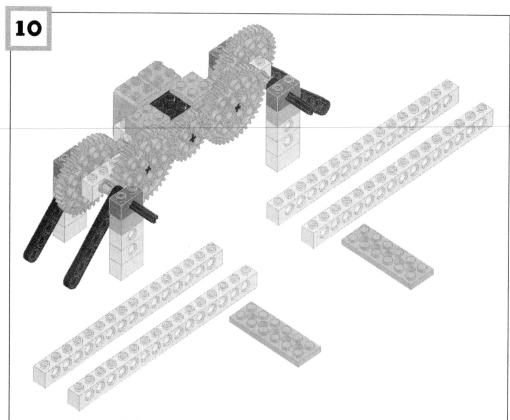

10

Slide the last hole in a 1x7 liftarm over each end of the two L8 cross axles and attach the four pillars outside these liftarms.

11

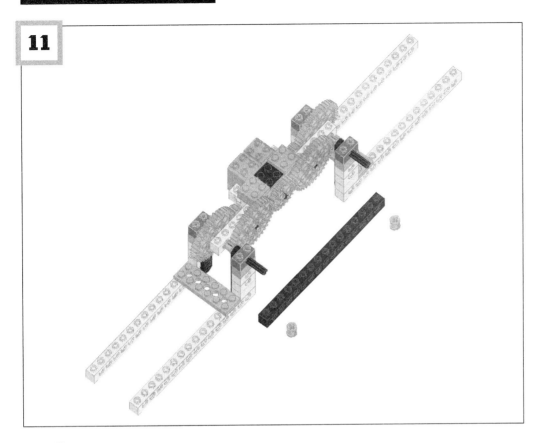

Attach a 1x16 beam to the bottom of each 1x2 pillar and secure them with 2x6 plates. Adjust the meshing of the gears so that each of the 1x16 beams is horizontal.

12

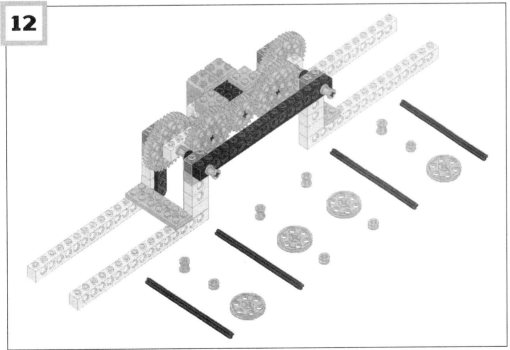

Slide a 1x16 beam onto the protruding cross axles and secure it with bushings.

13

Pass four L10 cross axles through the four 1x16 beams, then secure the axles' ends on the far side of the robot with bushings. Attach half bushings and pulleys on the ends of the cross axles on the opposite side of the robot; be sure that the half bushings are on the outside of the two middle cross axles and the pulleys are on the outside of the front and back cross axles. Also, make sure that the cross axles pass through the 1x7 liftarms.

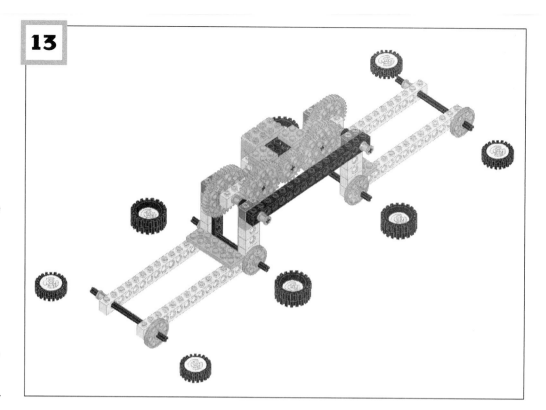

14

Attach tires to both ends of each of the four cross axles. Use medium tires for the four middle wheels and small tires for the front and rear wheels. If none of the wheels touch the ground, adjust the meshing of the gears until they do.

15

Prepare two motors to create separate driving forces for the front and rear ends of the frame. Attach two 1x6 plates to the bottom of one motor.

16

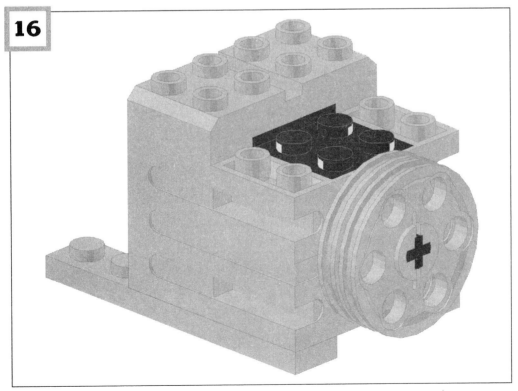

Now, slide two pulleys onto the cross axle of the motor. Repeat Steps 15 and 16 to build another motor.

17

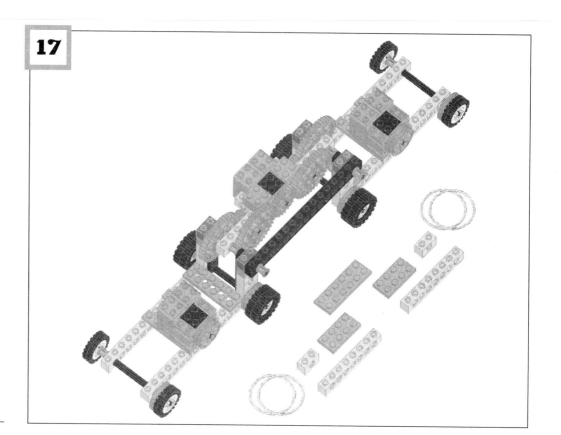

Attach both motors to the top of the 1x16 beams as shown, six pegs from the end of each beam.

18

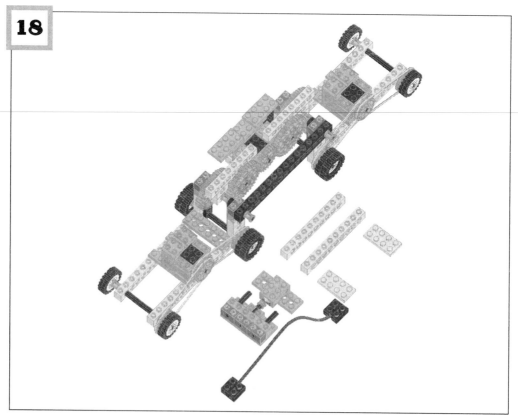

Stretch yellow rubber bands over the pulleys on the motors and the wheels. (The inner wheels should be attached to the inside pulleys, and the outer wheels to the outside pulleys.) Now, reinforce the top of the middle portion of the frame (in preparation for mounting the RCX) by attaching first a 1x2 beam, then a 1x8 beam on both sides of the top (center) motor. Also, mount one 2x6 plate, then two 2x4 plates onto the top motor.

19

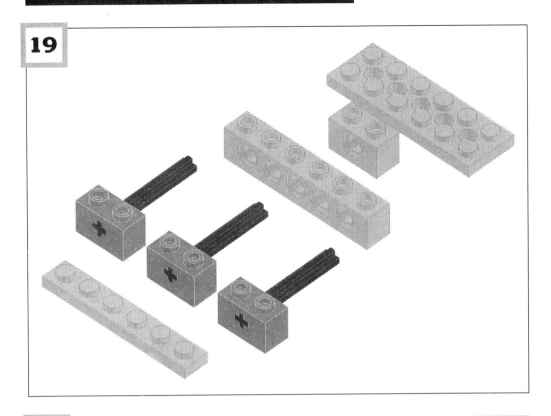

Gather the parts as shown. Insert three L4 cross axles into three 1x2 bricks with axle holes.

20

Connect the three 1x2 bricks with axle holes with a 1x6 plate to make the bumper. Now assemble the 1x6 and 1x2 beams and 2x6 plate, and attach this assembly to the bumper.

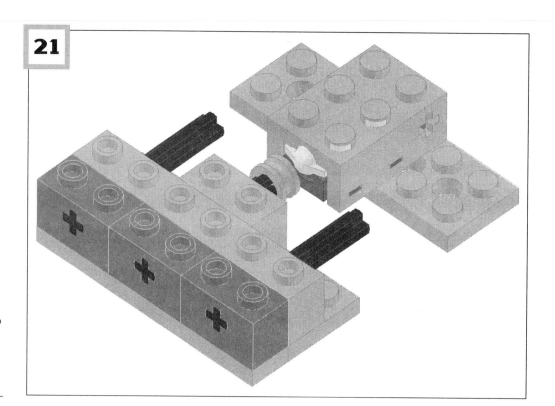

Attach a half bushing to
the end of the middle
cross axle. Now attach
a 2x6 plate to the bot-
tom of a touch sensor.

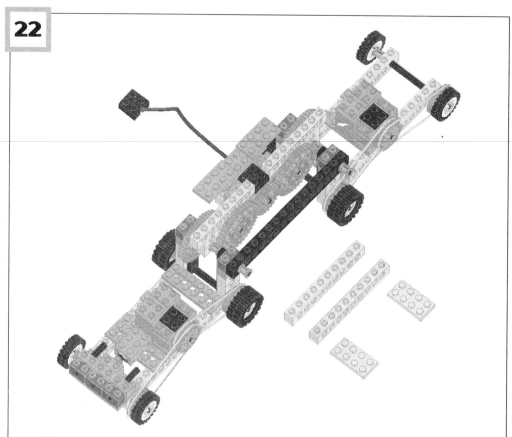

Attach the bumper and
touch sensor to the front
end of the frame and
adjust them so that when
the bumper touches an
obstacle the touch sensor
turns on. Attach a con-
necting lead to the
middle motor.

23

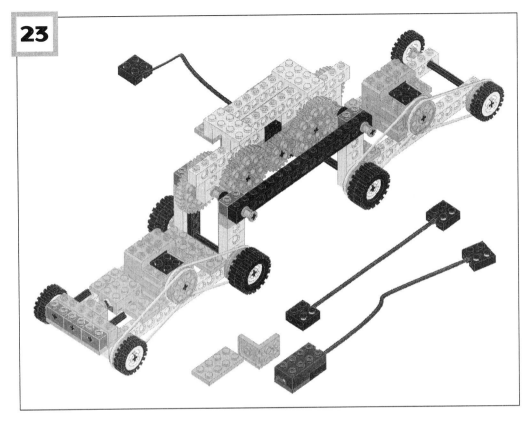

Make a flat surface on the middle portion of the frame as shown. Attach two 1x10 beams to the 1x8 beams and 2x4 plates you added in Step 17, then use two 2x4 plates to secure the 1x10 beams.

24

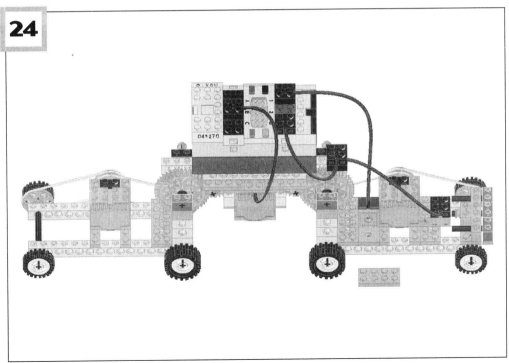

Attach a light sensor behind the front motor by attaching a 2x2–2x2 bracket to the 1x16 beams as shown (at right), then attaching the light sensor vertically to the bracket. Mount the RCX onto the support you created for it, then connect the touch sensor to input 3, the light sensor to input 1, and the middle motor to output B.

Secure the top of the light sensor bracket with a 2x4 plate. (The bracket has no studs, so attaching the 2x4 plate to the beams on both sides of it will keep the bracket from coming loose.) Attach a connecting lead from the front motor to output A and from the back motor to output C.

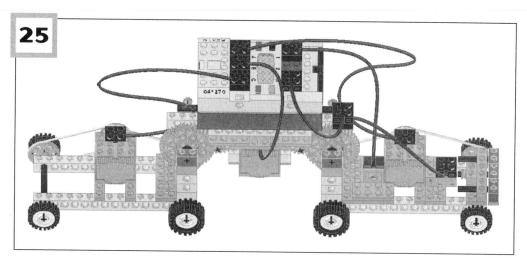

25

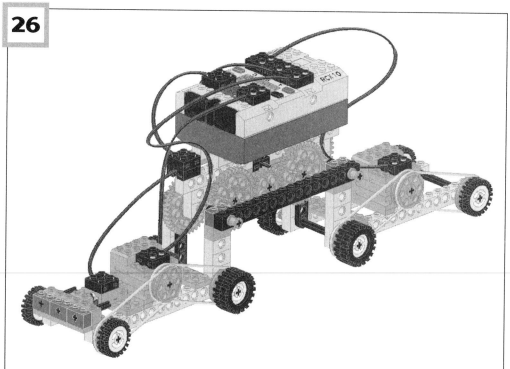

26

Here's another look at the completed model, offering a better view of the wiring.

PROGRAM

By controlling outputs A and C, we can rotate the eight tires to move the Climber forward and back. Also, by controlling output B, we can engage the middle motor to change the shape of the frame.

The following NQC program combines these operations so that the robot will climb over obstacles. (NQC, Not Quite C, is a programming language that lets you write programs on your computer to later download to the RCX. See http://www.enteract.com/ ~dbaum/nqc/ for more information.) Pressing the touch sensor performs the ascend operation; when the light sensor senses that the ground is moving away, the descend operation is performed.

```
#define LEVEL 33

int direction;
int count;

task main()
{
    SetSensor(SENSOR_1, SEN-
SOR_LIGHT);
    SetSensor(SENSOR_3, SEN-
SOR_TOUCH);

    Wait(100);
    SetPower(OUT_A, OUT_FULL);
    SetPower(OUT_B, OUT_FULL);
    SetPower(OUT_C, OUT_FULL);
    OnFwd(OUT_A);
    OnFwd(OUT_C);

    while(true){
        if(SENSOR_3 == 1){
            Off(OUT_A);
            Off(OUT_C);
            OnRev(OUT_B);
            Wait(40);
            Off(OUT_B);
            OnFwd(OUT_A);
            OnFwd(OUT_C);
            Wait(60);

            Off(OUT_A);
            Off(OUT_C);
            OnFwd(OUT_B);
            Wait(80);
            Off(OUT_B);
            OnFwd(OUT_A);
            OnFwd(OUT_C);

            Wait(60);

            Off(OUT_A);
            Off(OUT_C);
            OnRev(OUT_B);
            Wait(80);
            Off(OUT_B);
            OnFwd(OUT_A);
            OnFwd(OUT_C);
            Wait(60);

            Off(OUT_A);
            Off(OUT_C);
            OnFwd(OUT_B);
            Wait(80);
            Off(OUT_B);
            OnFwd(OUT_A);
            OnFwd(OUT_C);
            Wait(40);

            Off(OUT_A);
            Off(OUT_C);
            OnRev(OUT_B);
            Wait(10);

            Float(OUT_B);
            Wait(40);
            Off(OUT_B);
            OnFwd(OUT_A);
            OnFwd(OUT_C);
        }
        if(SENSOR_1 < LEVEL){
            Off(OUT_A);
            Off(OUT_C);
            OnFwd(OUT_B);
            Wait(40);
            Off(OUT_B);

            OnFwd(OUT_A);
            OnFwd(OUT_C);
            Wait(30);

            Off(OUT_A);
            Off(OUT_C);
            OnRev(OUT_B);
            Wait(80);
            Off(OUT_B);
            OnFwd(OUT_A);
            OnFwd(OUT_C);
            Wait(40);

            Off(OUT_A);
            Off(OUT_C);
            OnFwd(OUT_B);
            Wait(80);
            Off(OUT_B);
            OnFwd(OUT_A);
            OnFwd(OUT_C);
            Wait(40);

            Off(OUT_A);
            Off(OUT_C);
            OnRev(OUT_B);
            Wait(10);

            Float(OUT_B);
            Wait(40);
            Off(OUT_B);
            OnFwd(OUT_A);
            OnFwd(OUT_C);

        }
    }
}
```

TIP

When electricity flows to a motor, the motor rotates. But what if we turn the motor manually? We generate electricity, of course. Therefore, if we connect a motor to a fiber optic light unit with a lead and turn the motor's cross axle manually, the red LED of the fiber optic light unit should shine. (If it does not shine, try turning the cross axle in the center of the fiber optic light unit a little, because some orientations of the fiber optic light unit allow it to shine and others do not.)

LEGO light bricks

The LEGO light brick lights up when connected to a battery. If we connect a light brick to a motor, we instantly produce a hand-powered flashlight.

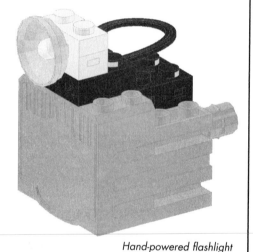

Hand-powered flashlight

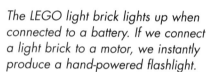
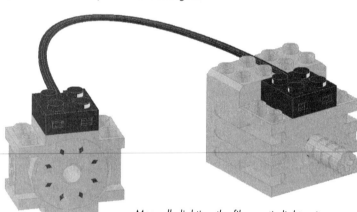

Manually lighting the fiber optic light unit

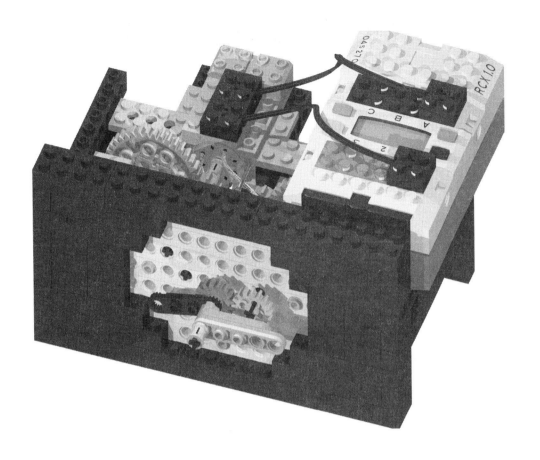

MODEL 8: LEGO CLOCK

The LEGO MINDSTORMS set contains various types of gears you can use to build a clock. Before you start, however, consider the ratios of the revolutions of a clock's hour, minute, and second hands: The ratio of the hour hand to the minute hand is 1:12, and the minute hand to the second hand is 1:60.

We'll give our clock a traditional hour hand and minute hand—then program a fiber optic light to shine sequentially to mark the passage of seconds.

The program for this LEGO clock causes the motor axle to make one revolution each minute. This would be simpler if we had a pre-built angle sensor, but that part is difficult to obtain. We also could, of course, build a mechanism that uses a touch sensor to check the axle's position, but instead we'll use the light sensor to measure the motor's rotation angle.

The fiber optic light's primary purpose is to emit light through its connected fibers, but we can also use it to create an angle sensor. The fiber optic light unit has a light emitting diode (LED) in its center and, when the unit rotates, fibers placed on the eight holes of the light unit will shine sequentially. Whereas an RCX *output* would normally be connected to the fiber optic light unit, we'll connect an RCX *input*, so that when the RCX input is set to light sensor mode it will make the fiber optic light shine.

HOW IT ALL WORKS

When the LED is shining, the RCX reads the input value as non-zero; when it turns off and rotates to the adjacent hole, this value returns to zero. We want to move the motor the distance between two holes on the light unit enough times to make a complete rotation (360° degrees) in one minute; thus, we need to move it eight times, once for every space between holes, every minute. The distance between holes is 45° (360° ÷ 8 holes = 45°). Our program must therefore rotate the motor 45° degrees every 7.5 seconds (1 minute = 60 seconds; 60 ÷ 8 = 7.5). We can achieve this by calling, once every 7.5 seconds, a program that rotates the motor during the time the light unit rotates between two holes (i.e., 45°).

PARTS LIST		
TYPE	**SIZE**	**QUANTITY**
Brick	1x2	18
Brick	1x12	2
Brick	2x2	2
Brick	2x4	6
Brick	2x6	8
Brick	2x8	4

PARTS LIST (CONT.)

TYPE	SIZE	QUANTITY
Beam	1x4	1
Beam	1x6	6
Beam	1x8	1
Beam	1x10	1
Beam	1x12	2
Beam	1x16	6
Liftarm	1x4	1
Liftarm	1x5	1
Plate	1x4	3
Plate	1x2 with rail	4
Plate	2x4	6
Plate	2x6	1
Plate	2x10	2
TECHNIC plate with holes	1x4 rounded	1
Cross axle	3	2
Cross axle	4	4
Cross axle	6	1
Cross axle	8	2
Gear	8-tooth	7
Gear	16-tooth	6
Gear	24-tooth	2
Gear	40-tooth	1
Gear	16-tooth clutch	1
TECHNIC pin	1/2	6
Half bushing		1
Bushing		5
Motor		1
Connecting lead	Short	2
Fiber optic light unit		1
Fiber		4

BUILD THE MOTOR HOUSING

1

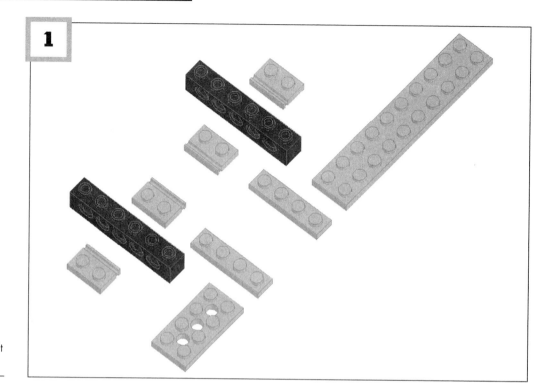

Gather the pieces for the base of the motor and fiber optic light unit assembly as shown.

2

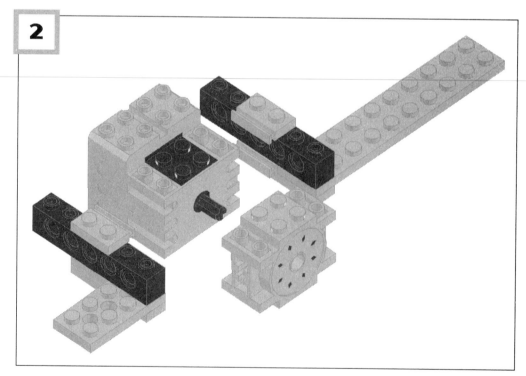

Center 1x2 plates with rails on the top and bottom of two 1x6 beams as shown. Attach a 2x4 plate to the bottom of one beam and a 2x10 plate to the bottom of the other; reinforce both with 1x4 plates.

3

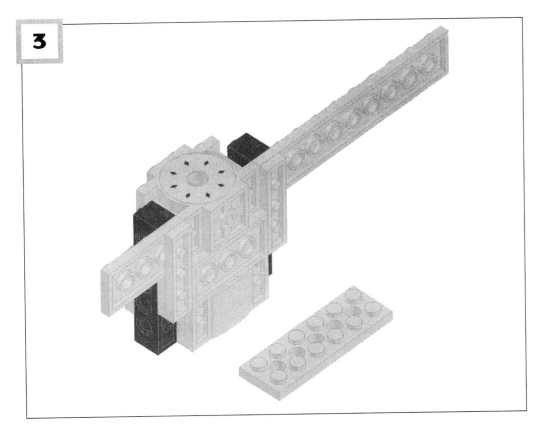

Insert the motor's axle into the fiber optic light unit. Slide the plates with rails onto the motor's grooves.

4

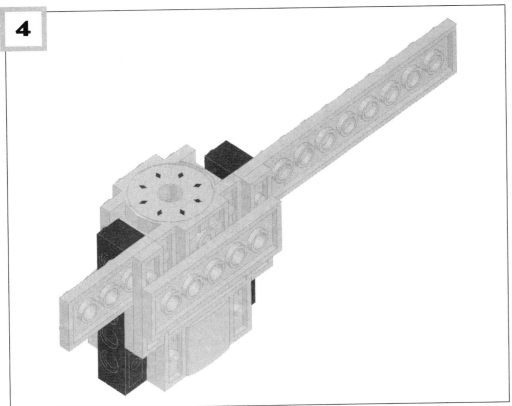

Secure the motor and fiber optic light unit assembly from the bottom with a 2x6 plate.

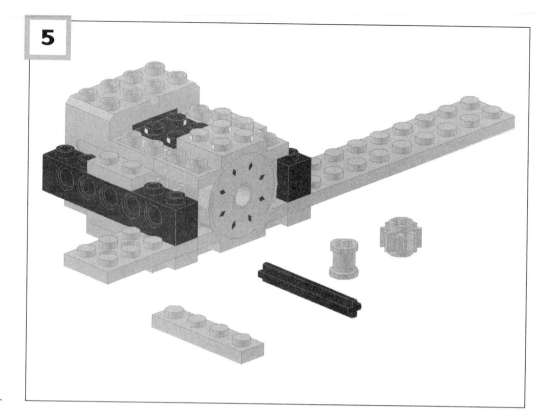

5

Gather an L4 axle, a bushing, a 1x4 plate, and an 8-tooth gear.

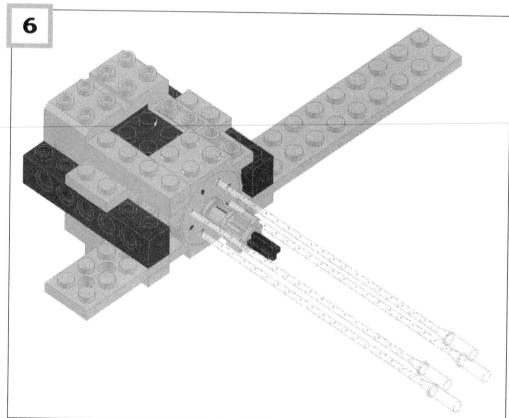

6

Slide the bushing and 8-tooth gear onto the L4 axle and insert the axle into the fiber optic light unit. Use the 1x4 plate to secure one side of the motor and light unit. Insert four fibers into the fiber optic light unit.

7

Mesh the small teeth of a 16-tooth clutch gear (the hole in the center enables it to coast) with the teeth on a 1x4 TECHNIC plate with holes to create the hour hand.

8

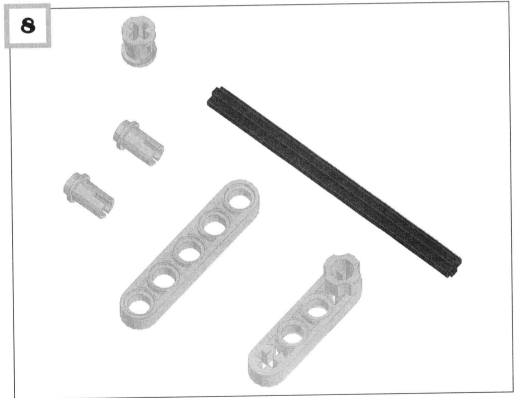

Gather the parts for the minute hand.

9

To build the minute hand, connect the 1x4 liftarm to the 1x5 liftarm with 1/2 TECHNIC pins. Pass the L8 axle through the minute hand and place a bushing on the end of the axle.

BUILD THE GEAR ASSEMBLY

10

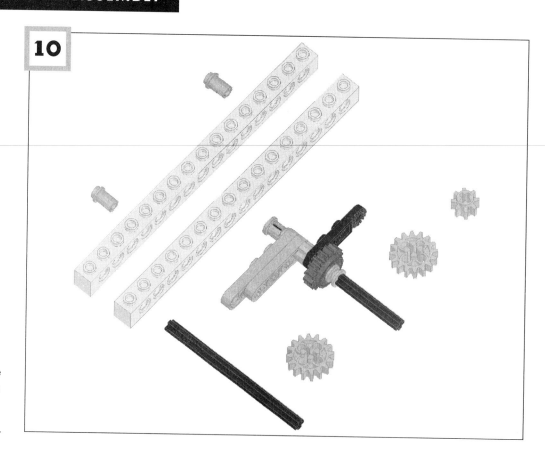

Pass the L8 axle with the minute hand through the hour hand and secure it with a half bushing.

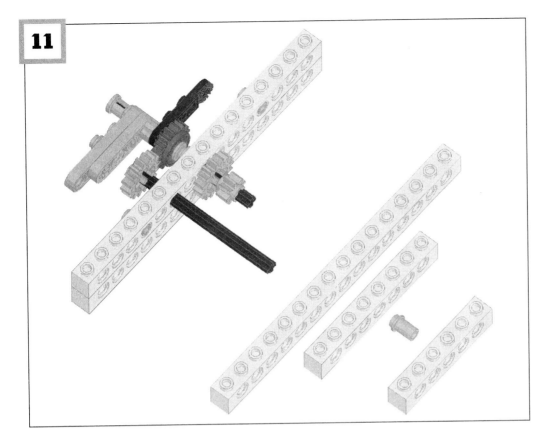

11

Stack two 1x16 beams and pass the L8 axle with hands through the center hole of the top beam. Attach a 16-tooth gear and then an 8-tooth gear to the axle. Slide another L8 axle through the same beam and attach a 16-tooth gear (see figure) to the axle so that the gear's teeth mesh with the clutch gear of the little hand. Insert 1/2 TECHNIC pins into the same 1x16 beam as shown.

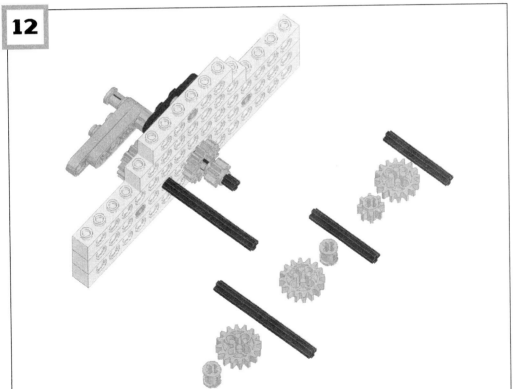

12

Stack 1x16, 1x8, and 1x6 beams on top of the 1x16 beam. Insert a 1/2 TECHNIC pin in the middle hole of the top beam.

13

Gather three 16-tooth
gears and three axles
(one L6 and two L4s) as
shown in the figure.
Pass the L6 axle through
one gear and slide on a
bushing; pass an L4
axle through another
gear and then slide on
a bushing; pass an L4
axle through the third
gear and then slide on
an 8-tooth gear as
shown in the figure.

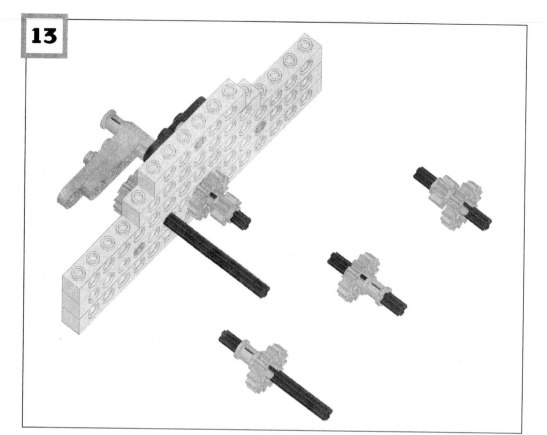

14

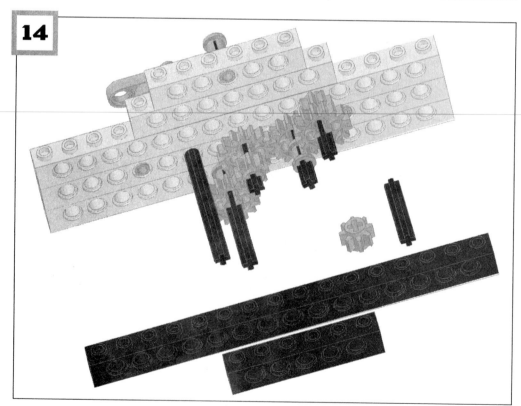

Insert the three gears
into the beams as
shown.

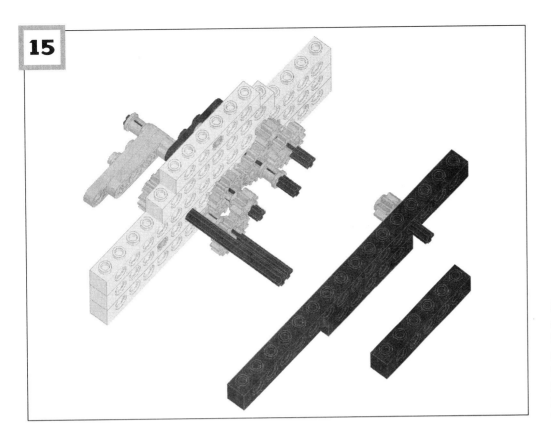

Stack a 1x16 beam and 1x6 beam, slide an 8-tooth gear onto an L3 axle, and insert the axle into the fourth hole from the right of the 1x16 beam.

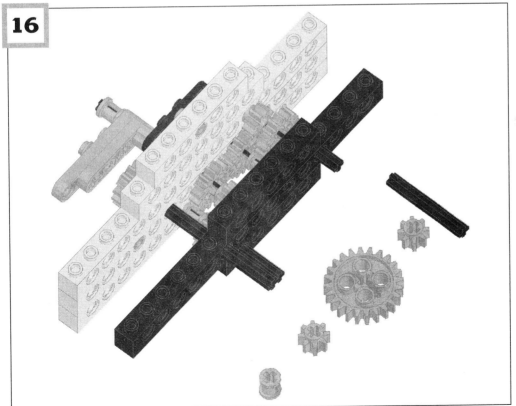

Add another 1x6 beam to this assembly and slide it onto the clock unit so that the 8-tooth gear meshes with the other gears assembled so far. (The 1x16 beams in both structures should line up exactly.)

17

Now attach the following pieces to the three axles that protrude from this second assembly: a bushing to the left axle, an 8-tooth gear to the middle axle, and a 24-tooth gear to the right axle. Also, attach an 8-tooth gear to an L4 axle and insert it in the beam (seen on the far right in the figure) so that the 8-tooth gear meshes with the 24-tooth gear.

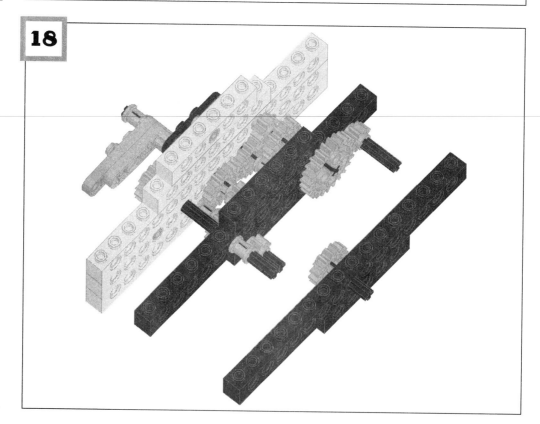

18

Combine 1x16 and 1x6 beams, then slide a 16-tooth gear onto an L3 axle, and pass the axle through the middle hole of the 1x16 beam so that the 16-tooth gear will mesh with the 8-tooth gear.

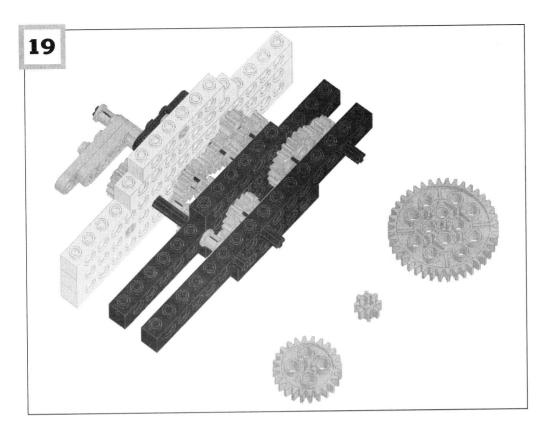

Slide this assembly onto the clock unit so that the gears mesh (the 1x16 beams should line up).

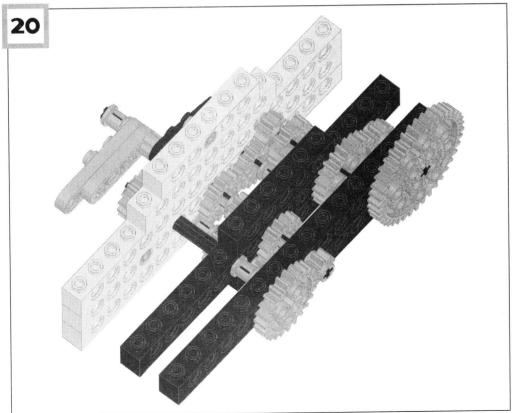

From left to right, attach a 24-tooth gear, 8-tooth gear, and 40-tooth gear to the axles that are protruding from the 1x16 beam.

21

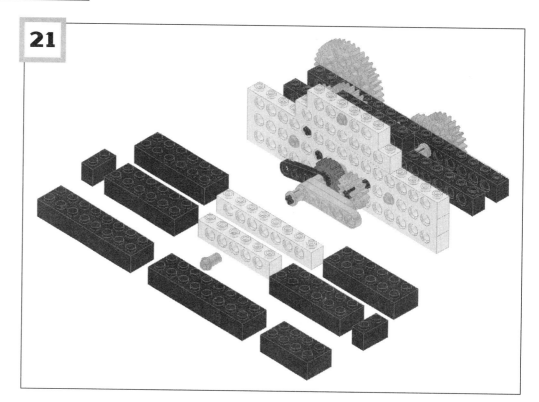

Gather the pieces shown in the figure to build the bottom of the clock.

22

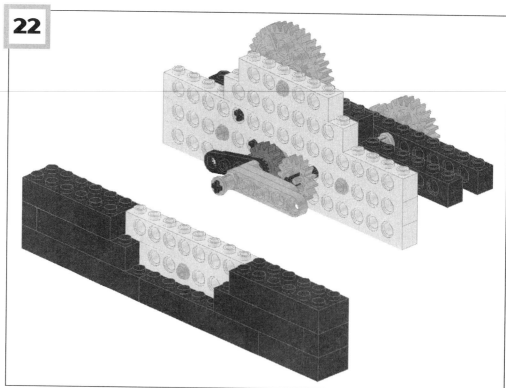

Use 1x8 and 1x6 beams and 1x2, 2x8, 2x6, and 2x4 bricks to build a foundation as shown. Insert a TECH-NIC pin into the 1x6 beam as shown.

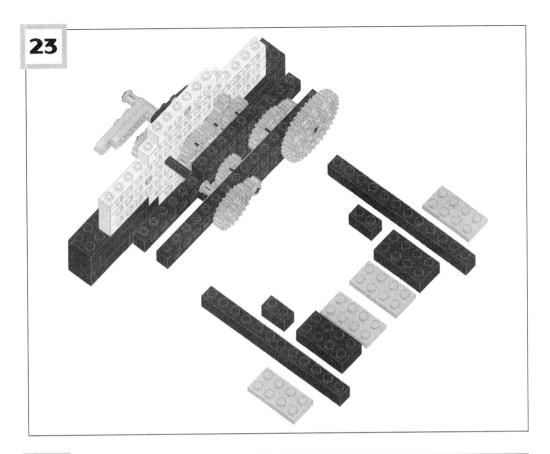

Mount the gear assembly (the beams with the hands attached) on the base you just created.

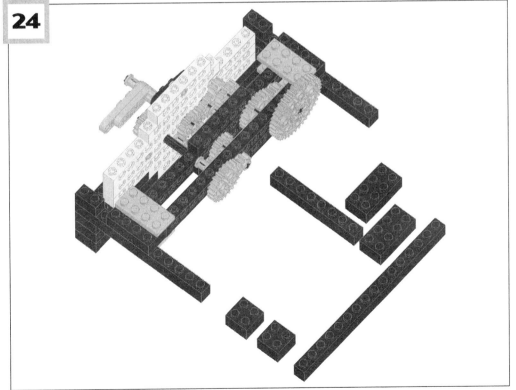

Secure two 1x12 bricks so that they extend away from the clock's face, as shown. Mount a 2x4 brick on top of each 1x12 brick and, using 2x4 plates, secure this frame assembly to the two 1x16 beams that support the gear assembly.

25

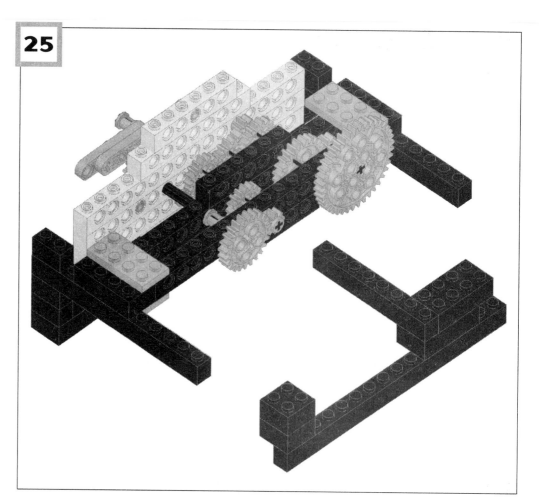

Build feet to support the clock by combining bricks as shown.

26

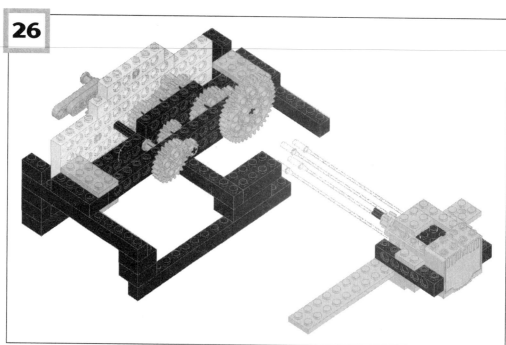

Attach the feet to the body to make the entire unit stand.

27

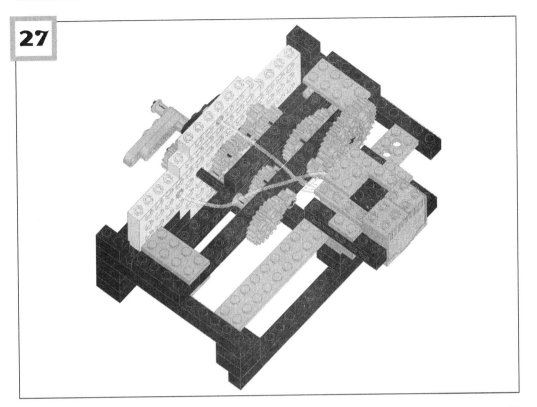

Attach the motor so that its 8-tooth gear meshes with the 40-tooth gear, then insert the four fibers into the 1/2 TECHNIC pins located at the four clock positions (12:00, 3:00, 6:00, and 9:00). (Because the motor axle rotates in the opposite direction from the axle of the big and little hands, two of the four fibers must be crossed.)

28

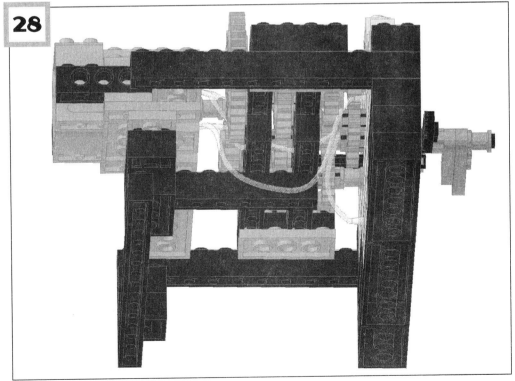

Adjust the fibers so that they do not contact the gears. Note how the fibers are crossed in this figure, which shows the model from below.

29

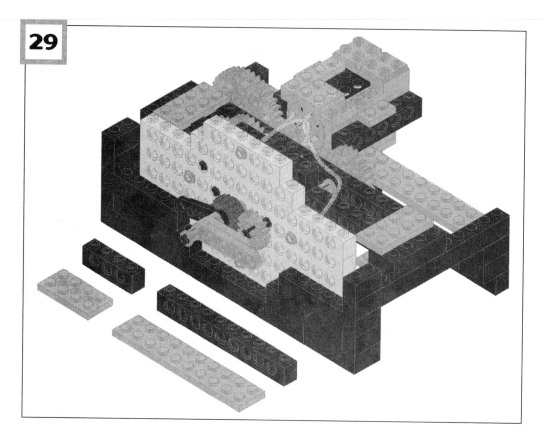

Gather the pieces shown to reinforce the assembly supporting the motor.

30

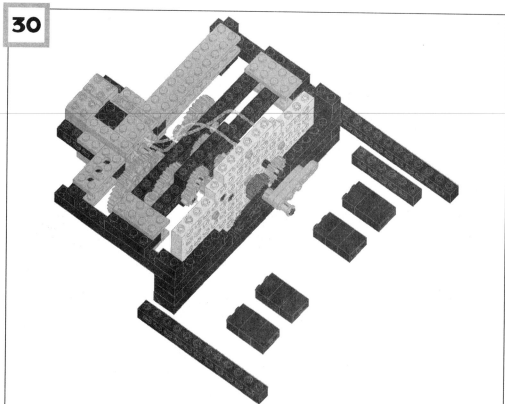

Use the 1x10 and 1x4 beams and 2x10 and 2x4 plates to reinforce both sides of the motor, as shown at upper left. Gather the pieces for the sides and face.

31

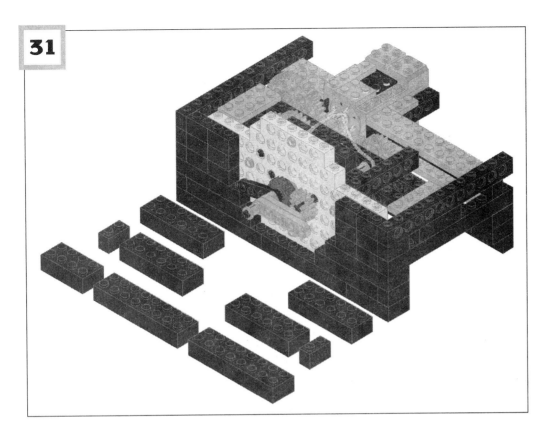

Stack 1x2 bricks to fill out the face; reinforce the sides with the 1x12 beams.

32

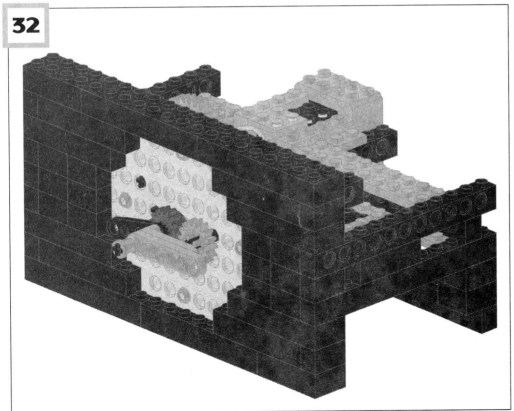

Stack bricks to complete the face of the clock.

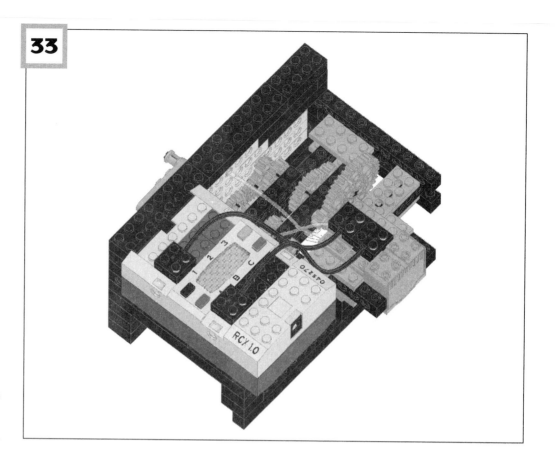

Mount the RCX.
Connect output A to the
motor and input 1 to the
fiber optic light unit to
complete the Clock.

PROGRAM
PROGRAM WRITTEN IN NQC

The main program, shown below, starts the task named second every 7.5 seconds by checking the rotation angle of the motor. The program uses the fact that the value of input 1 is non-zero when the LED of the fiber optic unit is shining, and zero when it is off.

The first while statement rotates the motor until the LED turns off (the input value is zero). The second while statement rotates the motor until the LED turns on again (the input value is non-zero). Because the fiber optic unit's LED shines at eight positions, this task is called once every 7.5 seconds, which is equal to 60 seconds divided by 8, causing the motor to complete one revolution every minute.

```
task main()
{
    SetSensor(SENSOR_1, SENSOR_LIGHT);
    SetPower(OUT_A, 1);
    start second;
    do{
        Wait(750);
        stop second;
```

```
        start second;
    }while(true);
}

task second()
{
    OnFwd(OUT_A);
    do{
    }while(SENSOR_1 != 0);
    do{
    }while(SENSOR_1 == 0}
    Off(OUT_A);
}
```

PROGRAM ACCORDING TO RCX CODE

The CLOCK program calls the my-commands (Figure 8-1) named lightoff and lighton
every 7.5 seconds. The lightoff my-command rotates the motor until the light emitting
diode goes off. The lighton my-command rotates the motor until the light emitting diode
lights up. Therefore, by calling these two my-commands, this program can rotate the axle
of the motor 45° every 7.5 seconds—and this causes the motor to complete one revolution every minute.

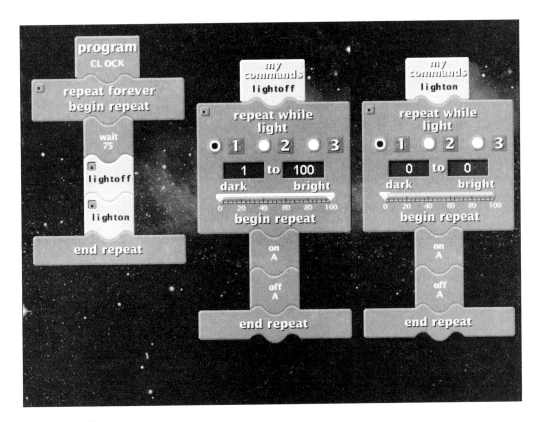

Figure 8-1: Clock program

TIP

The Tip on page 112 describes how to use the motor as an electric generator. We can use this same generator to build an analog remote control.

Several remote control robots in the LEGO MINDSTORMS manual can be controlled by a touch sensor: A handy switch connected to the robot by a lead. Pressing the touch sensor acts as the on and off switch. Here, however, we will use a motor as a remote control.

If you connect two motors with a lead and turn one motor's cross axles, you'll see that the other motor turns by exactly the same amount. This is the theory behind an analog remote control. Not only is this an easy way to build a remote control, but it's a good way to test a robot's movements, by turning the motor a little at a time.

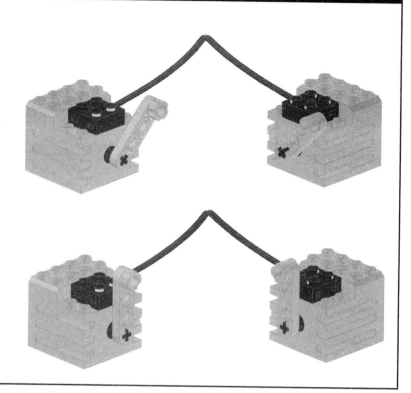

MODEL 9: PNEUMATIC ENGINE CAR

Can something other than a motor be used to move a robot? The short answer is yes: We can use the pneumatic pistons found in the LEGO TECHNIC series. These pistons, cylinders filled with compressed air, can drive the shovel of a bulldozer or steam shovel (a job handled by hydraulic pistons in the real world).

We'll use these TECHNIC pneumatic cylinders to build a pneumatic engine, the force of which will move a car.

The principles underlying a pneumatic engine are the same as that of a steam locomotive. As shown in Figure 9-1, in position (a), air forced into the left side of the piston generates a clockwise rotating force. In position (c), air forced into the right side of the piston also generates a clockwise rotating force. The midway points (b) and (d) are dead points, where the force of the piston does not contribute to rotation.

To eliminate the dead points, (b) and (d), we'll use two pneumatic cylinders and off-set their phase by 90° so that the engine will rotate smoothly.

Now let's try building a two-piston pneumatic engine.

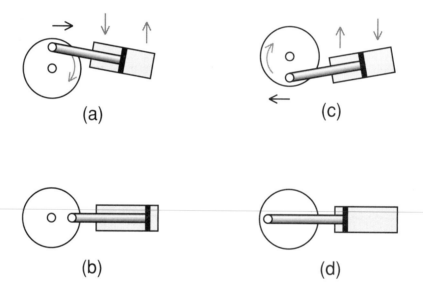

Figure 9-1: *The theory behind a steam locomotive*

PARTS LIST

TYPE	SIZE	QUANTITY
Brick	1x2	4
Brick	2x2	9
Brick	2x4	2
Beam	1x2	9

PARTS LIST (CONT.)

TYPE	SIZE	QUANTITY
Beam	1x6	2
Beam	1x8	4
Beam	1x10	2
Beam	1x16	3
Liftarm	1x4	2
Liftarm	1x5	2
Plate	1x2	1
Plate	1x3	1
Plate	1x6	1
Plate	2x4	2
Plate	2x6	3
Plate	2x10	4
Gear	24-tooth	1
Gear	40-tooth	2
Pulley	Medium	3
Pulley	Large	2
Cross axle	2	2
Cross axle	3	5
Cross axle	4	2
Cross axle	6	2
Cross axle	10	1
Cross axle	12	3
Axle joiner		4
Friction pin	2	4
TECHNIC pin	1/2	2
TECHNIC pin	3/4	3
Half bushing		5
Bushing		12
Angle connector	#1	2
Motor		1
Connecting lead	Short	2

PARTS LIST *(CONT.)*

TYPE	SIZE	QUANTITY
Wheel hub	Extra large	4
Rubber tire	Extra large	4
T connector		1
Pneumatic cylinder	Small	1
Pneumatic cylinder	Large	2
Pneumatic valve		2
Air tank		1
Pneumatic tubing	2 cm	1
Pneumatic tubing	3 cm	1
Pneumatic tubing	5 cm	1
Pneumatic tubing	9 cm	1
Pneumatic tubing	15 cm	4

ASSEMBLY PROCEDURE

BUILD THE COMPRESSOR

1

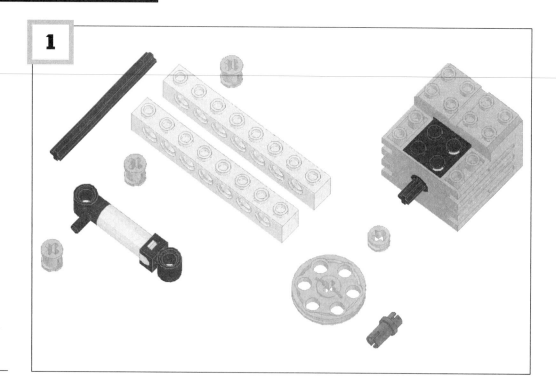

First, gather the pieces shown to build the motor.

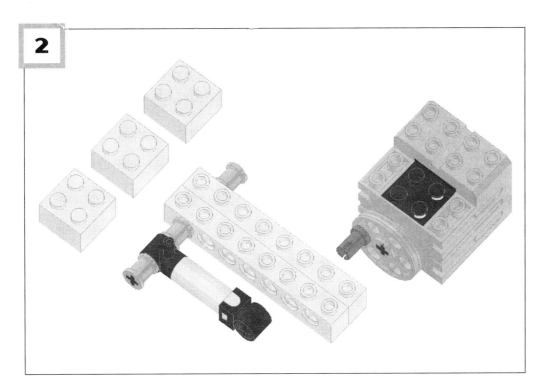

2

Align two 1x8 beams, pass an L6 cross axle through them, and attach a small pneumatic cylinder to make a compressor. Attach bushings to the cross axle as shown, then slide a half bushing onto the motor cross axle and attach a pulley. Attach the 3/4 TECHNIC pin to the pulley.

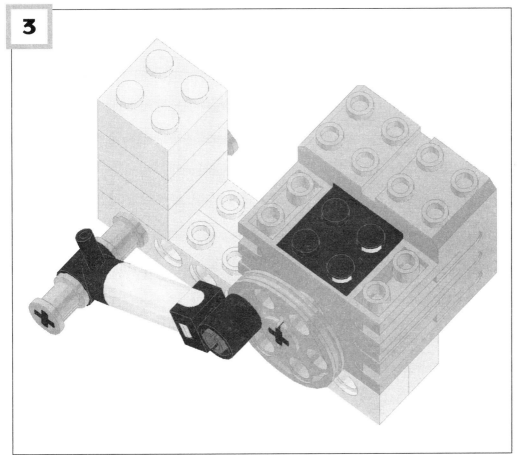

3

Mount the motor on the 1x8 beams and slide the opposite end of the piston over the 3/4 TECHNIC pin on the motor's pulley. Attach the 2x2 bricks.

4

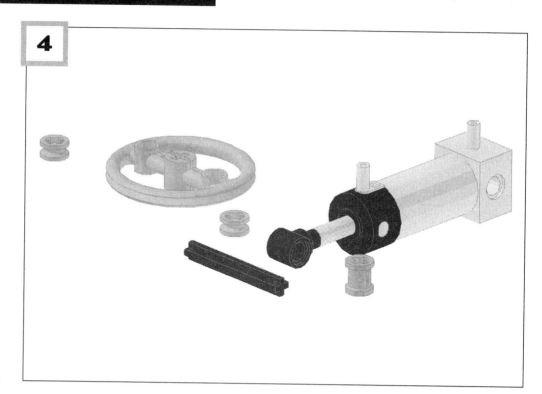

Assemble the pieces for the pneumatic piston unit, which will play the main role in the pneumatic engine.

5

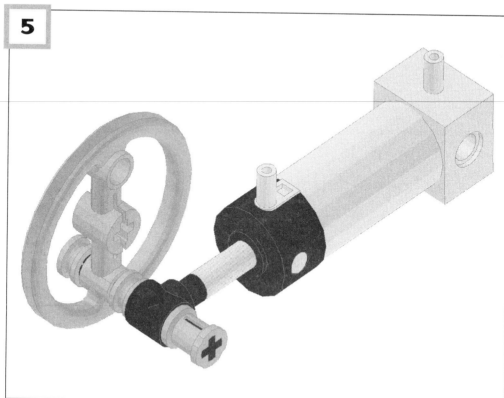

Attach the large pneumatic cylinder to a large pulley with an L4 cross axle, passing the cross axle through one of the pulley's side holes and securing both ends of the axle with half bushings. Next, build another unit that is identical but symmetrically reversed.

6

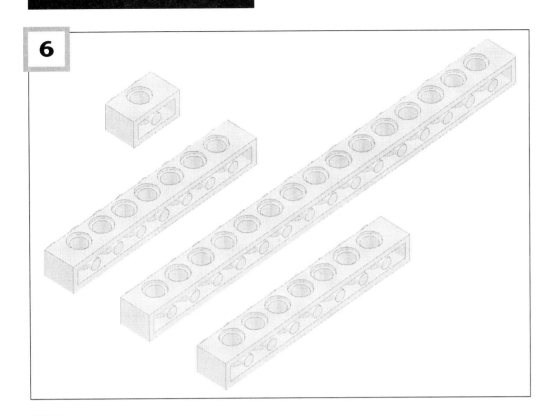

Gather one 1x16, two 1x8, and one 1x2 beams.

7

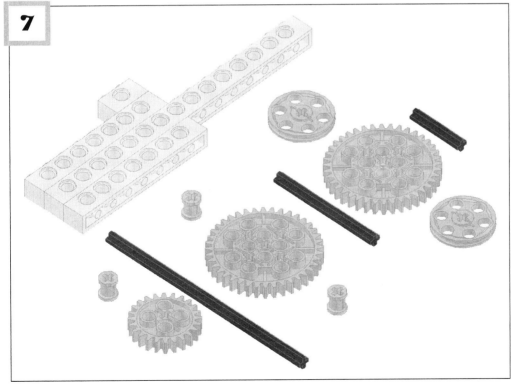

Stack the beams, then gather L12, L6, and L3 cross axles (one each) and the other parts shown.

Attach a 24-tooth gear to the L12 cross axle and a 40-tooth gear to the L6 cross axle, then insert the L3 cross axle through one pulley and then a 40-tooth gear. Next, insert the cross axle through the sixth hole of the 1x16 beam. Mount the second medium-sized pulley on the protruding end of the L3 cross axle (one pulley goes on the outside of the 1x16 beam, the other on the opposite side of the gear). Insert the other axles through the beams as shown.

8

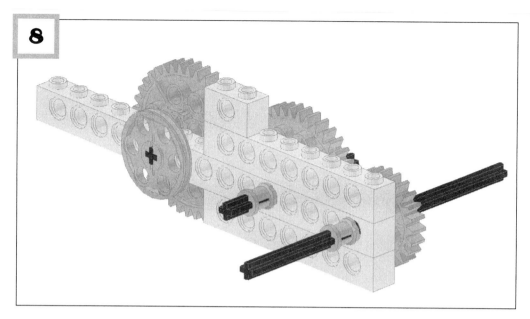

Check the orientation of the two medium-sized pulleys to be sure they are offset by 90°, as shown in the figure.

9

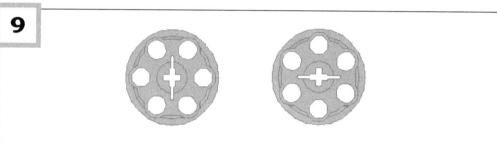

10

Assemble the parts shown to build a base for the pneumatic valves.

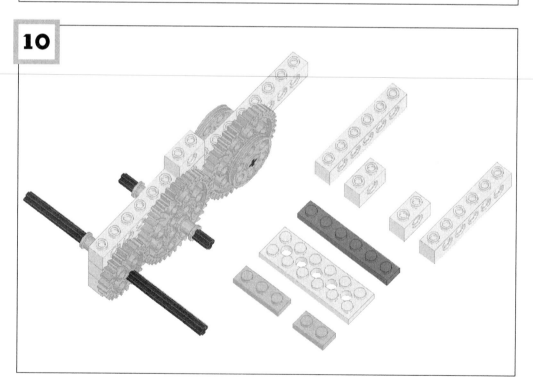

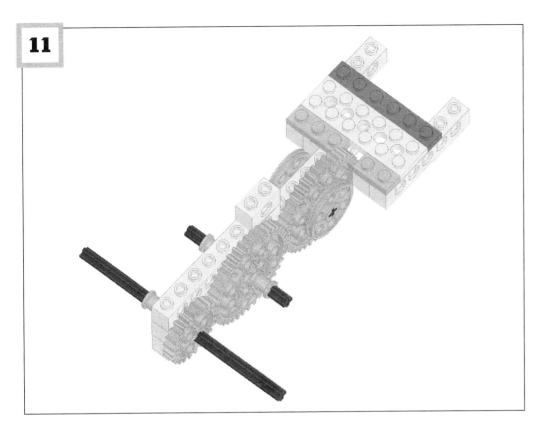

11

Secure the 1x6 beams
with a 2x6 plate; attach
the 1x2 beams next to
the 1x6 beams. Add a
1x6 plate for additional
support and attach a
1x2 and 1x3 plate to
the top of the 1x8
beams, allowing space
for the 40-tooth gear.
(This figure shows the
unit viewed from
above.)

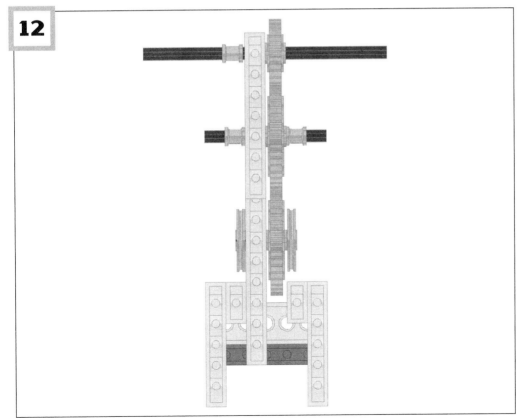

12

Confirm the positions of
the 1x2 beams. (This
figure shows the unit
from below.)

13

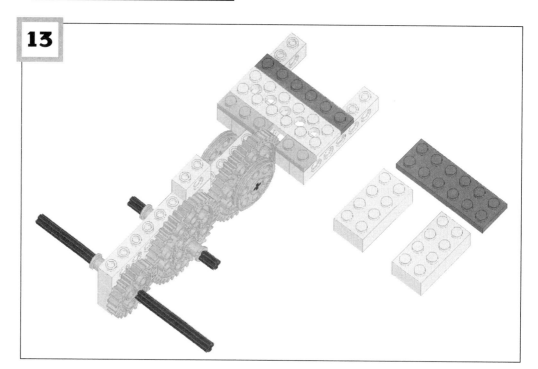

Gather the parts shown to add the base for mounting the pneumatic valves.

14

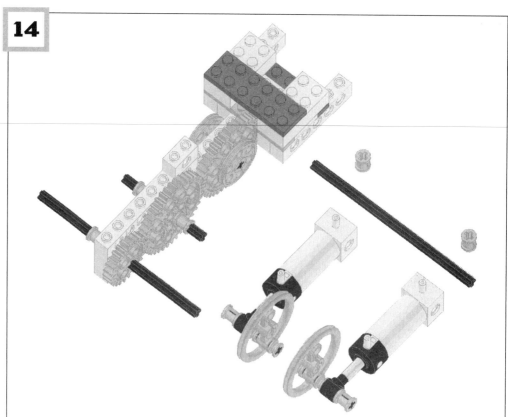

Place the 2x4 bricks on top of the base you built in Steps 10 through 12 and secure them with a 2x6 plate.

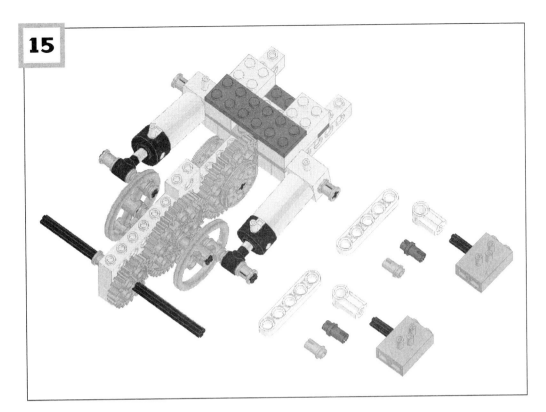

15

Attach the left and right pneumatic cylinders to the sides of the car's body on an L12 cross axle passed through the 1x6 beams. Confirm that the positions of the two large pulleys are offset by 90° (see Step 9).

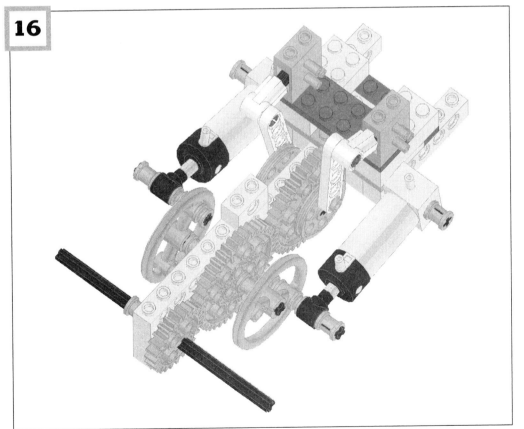

Attach a #1 angle connector to a pneumatic valve and use a 3/4 TECHNIC pin to connect this assembly to a 1x5 liftarm. Connect the opposite end of the liftarm to one of the medium-sized pulleys with a 1/2 TECHNIC pin, inserting the small end of the pin into the pulley. Repeat for the other side.

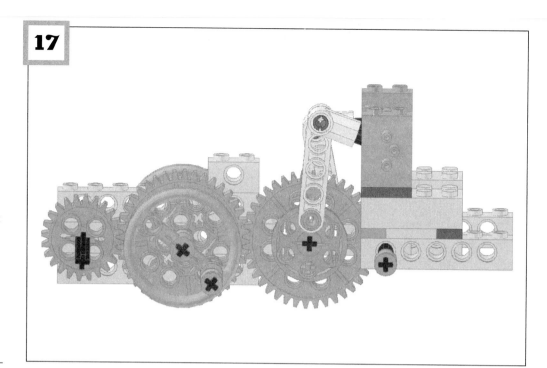

17

The relative positions of the large and medium pulleys are extremely important for proper operation of the pneumatic engine. Check that your car's pulleys are aligned as shown here.

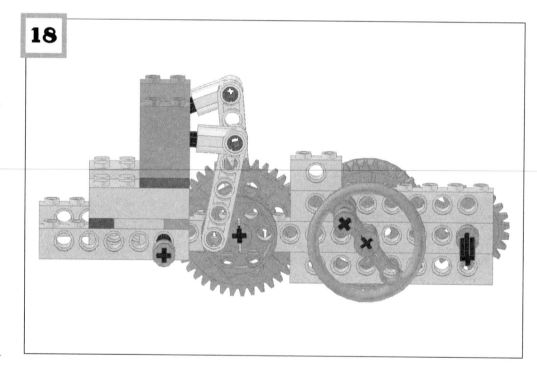

18

This figure shows the same assembly from the opposite side.

For the pneumatic engine to rotate smoothly, you must find the optimal positions between the pulleys, which can only be reached by adjusting the meshing of these two gears. (If you need to adjust the relative positions, be sure that the meshing of the two 40-tooth gears does not slip.)

19

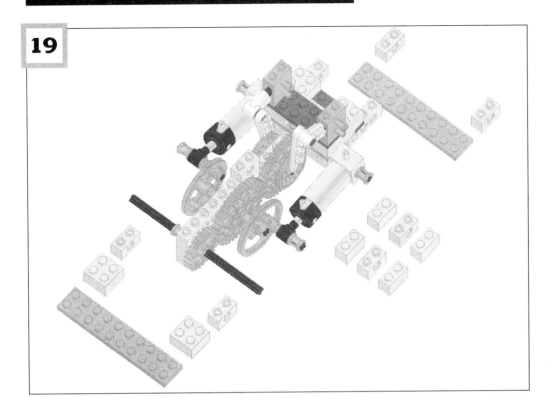

Gather the pieces for reinforcing the frame, as shown.

20

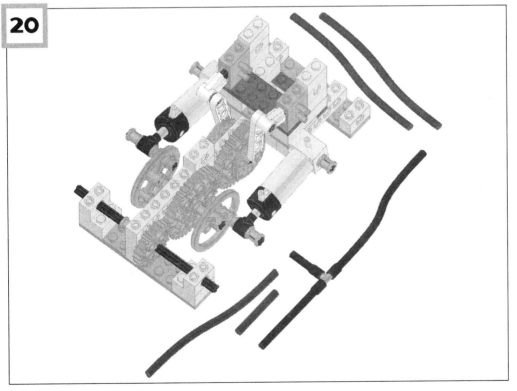

Reinforce the frame by attaching a 2x2 brick and a 1x2 beam on each end of a 2x10 plate, then attach this structure to the front of the car, passing the L12 axle through the 1x2 beams. For the back, attach 1x2 beams to each end of a 2x10 plate and attach the plate to the bottom of the car's 1x6 beams. Stack 1x2 bricks and beams near the valves as shown.

Connect pneumatic tubing of lengths 9 cm, 5 cm, and 2 cm to a T connector as shown, then gather three pieces of 15 cm tubing and one of 3 cm.

Connect the 9 cm pneumatic tubing to the middle nozzle of the right pneumatic valve (in the upper background of the figure), then connect the 2 cm piece to the middle nozzle of the left pneumatic valve (in the foreground of the figure).

Use the 3 cm pneumatic tubing to connect the bottom nozzle of the left pneumatic valve (foreground of the figure) to the base of the pneumatic cylinder next to it. Connect the top nozzle of the left pneumatic valve to the front nozzle of the left pneumatic cylinder with a 15 cm piece of pneumatic tubing.

Use another 15 cm pneumatic tube to connect the bottom nozzle of the right pneumatic valve (upper background of the figure) to the base of the right pneumatic cylinder, and another to connect the top nozzle of the right valve to the front nozzle of the right piston.

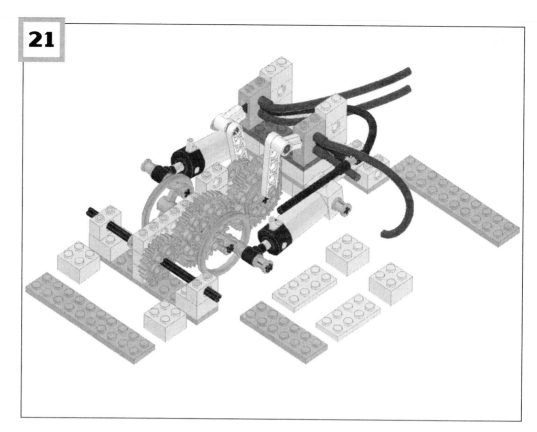

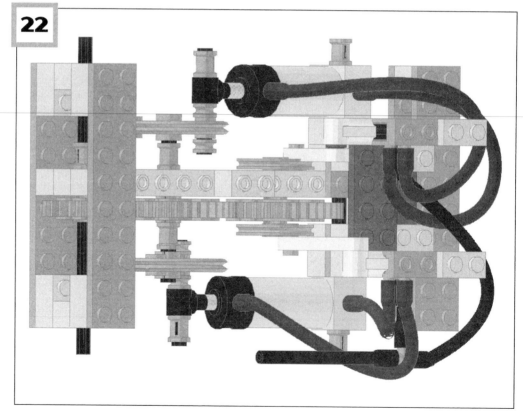

This figure shows the pneumatic tubing connections, viewed from above. If you have a hand-operated air pump, send compressed air into the 5 cm front pneumatic tubing attached to the T connector to confirm that the engine rotates. If the pistons are cold, the engine may be difficult to operate, so be sure to work in a warm room.

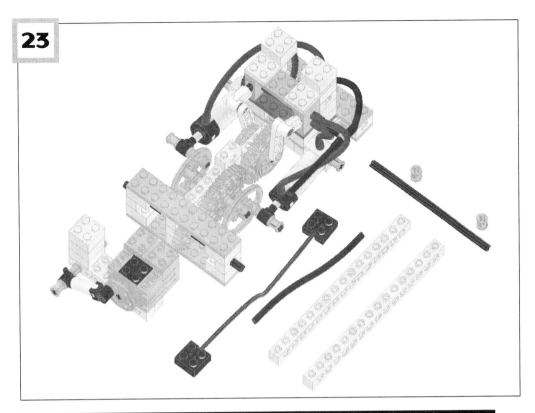

23

Reinforce the tops of the front wheel axle and the pneumatic valves with 2x2 bricks and plates. Reinforce the valves with the 2x6 plate, two 2x4 plates, and two 2x2 bricks; reinforce the back wheel axle.

ATTACH THE COMPRESSOR, AIR TANK, AND RCX

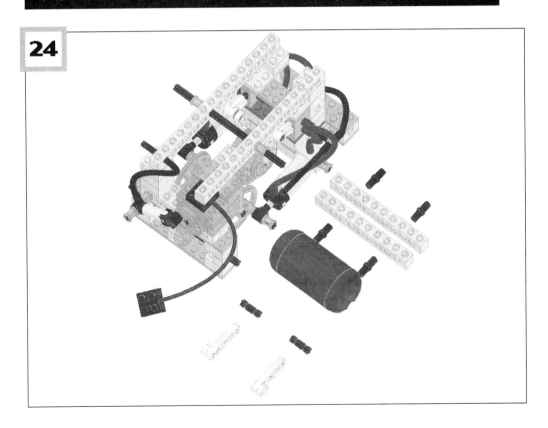

Attach the compressor unit and connect a 15 cm piece of pneumatic tubing to it. Also, attach two 1x16 beams along the top and slide the L10 axle through them as shown; attach the connecting lead to the motor.

Attach the air tank on
top of the central 1x2
beam and secure it on
both sides with 1x4 lif-
tarms; connect it to the
5 cm piece of tubing
coming from the T con-
nector and the 9 cm
tubing coming from the
compressor unit (at the
front of the car). Place
1x10 beams vertically
at the back side of the
assembly and secure
them with friction pins.

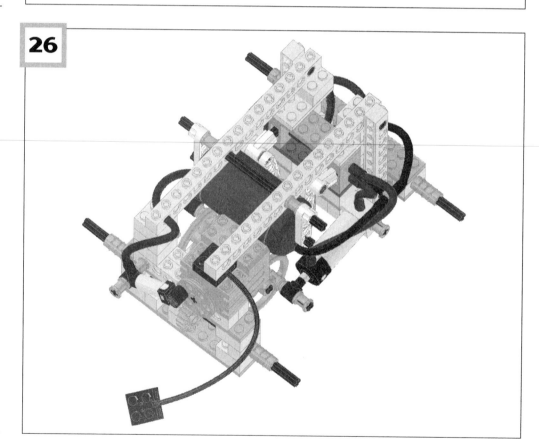

26

Connect axle joiners
to the front and back
wheel axles, then con-
nect L3 cross axles to
the axle joiners.

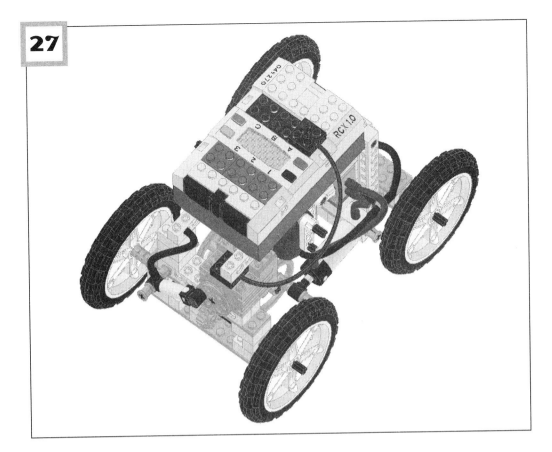

27

Finally, slide the wheels onto the cross axles, mount the RCX on top, and attach the connecting lead from the motor to output A to complete the car.

PROGRAM

This robot, which uses only output A, can move according to the default RCX program. Use the "Prog" button to select 1 and press "Run" to start the program.

Try building your own unique car by adding light or touch sensors.

MODEL 10: BEETLE

In this chapter we'll build a Beetle—a six-legged insect-like robot that can turn left or right. Although the Beetle is basically the same as the four-legged LEGOsaurus (Model 4), creating the ability to turn left or right is complicated, so adding this function makes it a more advanced robot.

The concept behind walking on six legs is nearly identical to that already presented for the four-legged LEGOsaurus. When the front right leg, middle left leg, and rear right leg are raised, the remaining legs must be touching the ground. Thus, the Beetle can walk forward and backward in a manner similar to the LEGOsaurus, as shown in Figure 10-1. The central black dot represents the robot's center of gravity. The four photographs in Figure 10-2 show the Beetle in action.

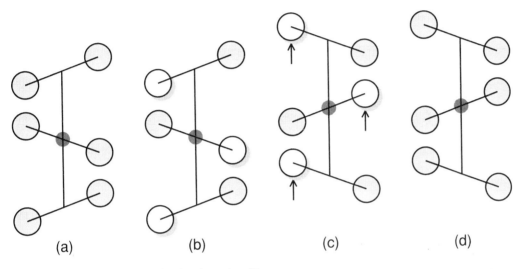

(a)	(b)	(c)	(d)

Figure 10-1: *The principles behind six-legged walking*

Turning, however, is a tougher proposition. We can make the Beetle turn left or right using two methods: One is to use two separate motors to move the three legs on the left and right sides, respectively. For example, if the left legs are moved while the right legs remain stationary, the robot will turn right.

This first method is a simple one, but it has a major disadvantage: Because the left and right legs are not synchronized (that is, when one side stops, the other would keep going), the robot's balance is unstable. For example, assume that the middle right leg is touching the ground, the front and rear right legs are raised, and the right legs are stopped so the beetle can begin a turn. As long as the left legs all remain on the ground, the robot is stable. However, once the front and rear left legs are raised to begin the next step (leaving only the middle left leg on the ground), the robot becomes unstable—because it is standing on only the two middle legs.

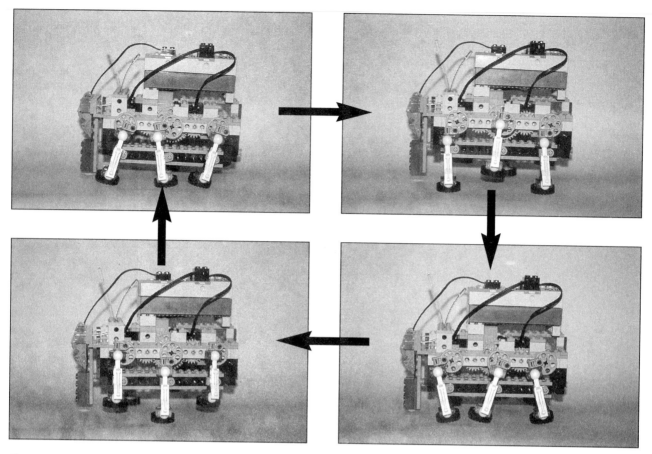

Figure 10-2: *The Beetle in action*

The second method solves this problem by varying the lengths of the strides of the left and right sides. To understand how this method works, consider what happens when a person walks along a road that turns to the right. Although the left and right feet are placed on the ground alternately, just as when walking straight, the right foot's stride shortens and the left foot's elongates as we round the curve. Similarly, even when the left and right legs are synchronized, a six-legged robot will turn right if its right-foot stride is shorter than its left.

The length of the robot's stride is determined by a fulcrum. We can use the fulcrum to change the robot's stride by changing its location, as shown in Figure 10-3. (The fulcrum is the gray rectangle with the dashed line in the center of each image; the length of stride is the span of the double-headed arrows.) When the fulcrum is lowered as in (a), the stride narrows because the leg is effectively made shorter (see the figure above of the robot—the perpendicular axle joiner on the leg acts as the fulcrum). When it is raised as in (b), the stride widens.

Let's tackle building a robot that turns left and right by varying its stride.

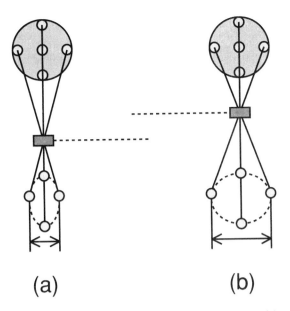

Figure 10-3: *Varying a robot's stride by moving the fulcrum*

PARTS LIST		
TYPE	**SIZE**	**QUANTITY**
Brick	1x2	9
Brick	2x2	2
Brick	2x4	1
Brick	2x6	1
Brick	2x2 round	2
Beam	1x2	18
Beam	1x4	5
Beam	1x6	7
Beam	1x8	8
Beam	1x16	2
Brick	1x2 with axle hole	3
Plate	1x2	13
Plate	1x2 with rail	7
Plate	1x4	6
Plate	1x6	1
Plate	1x8	5
Plate	1x10	2

PARTS LIST (CONT.)

TYPE	SIZE	QUANTITY
Plate	2x2 corner	1
Plate	2x2	2
Plate	2x4	5
Plate	2x6	6
Plate	2x8	7
Bracket	2x2–2x2	2
Liftarm	1x3	2
Liftarm	1x5	1
Liftarm	1x7	3
Gear	8-tooth	1
Gear	24-tooth	5
Gear	40-tooth	1
Pulley	Medium	6
Cross axle	8	7
Cross axle	10	3
Cross axle	12	1
Axle joiner		1
Friction pin	3	1
TECHNIC pin	3/4	11
TECHNIC pin	1	4
TECHNIC pin with axle		8
Bushing	1	8
Motor		2
Connecting lead	Short	3
Light sensor		1
Fiber optic light unit		1
Wheel hub		6
Rubber tire		6
SLIZER axle		6
SLIZER bearing		12
SLIZER head		1

ASSEMBLY PROCEDURE

BUILD THE MOTOR ASSEMBLY

1

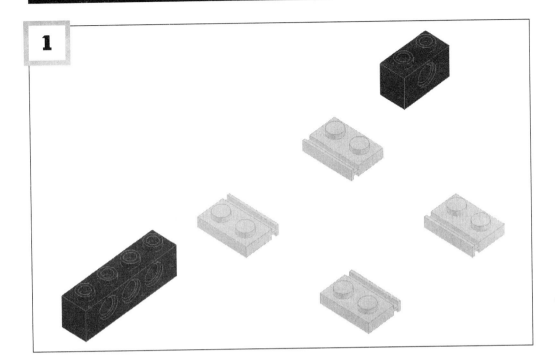

Collect the pieces shown to begin the motor assembly.

2

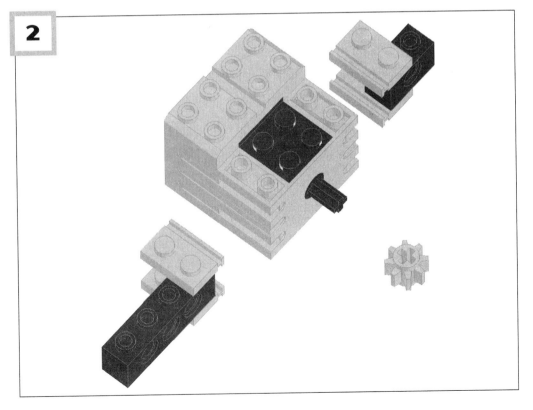

Attach two 1x2 plates with rails each to a 1x4 and a 1x2 beam, then slide an 8-tooth gear onto the motor's cross axle.

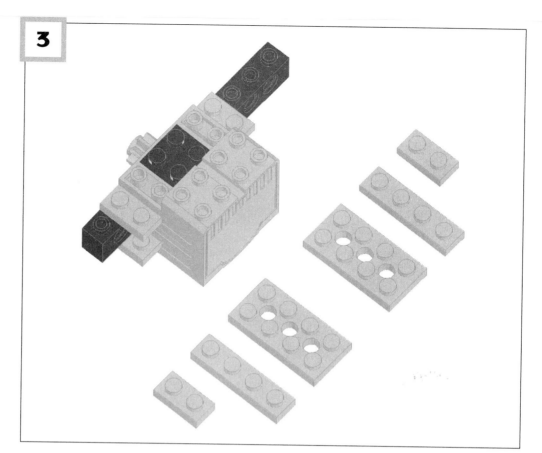

Insert the plates with
rails into the channels
on the sides of the
motor.

4

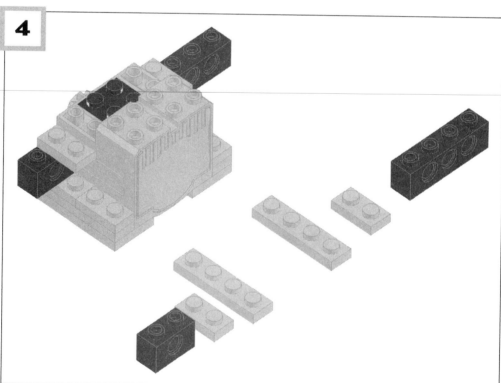

Stack 1x2, 1x4, and
2x4 plates and attach
them to the bottom of
the motor.

5

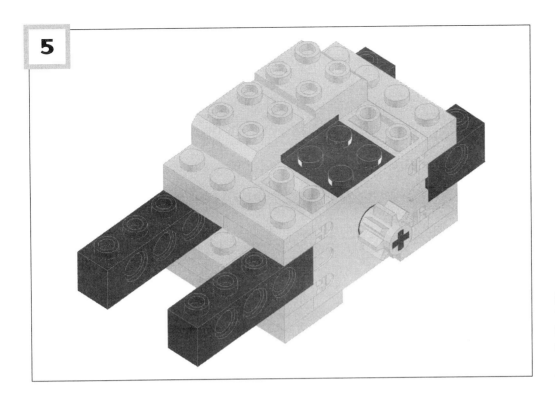

Mount a 1x2 and a 1x4 beam and secure them on top of the motor with 1x2 and 1x4 plates, as shown.

6

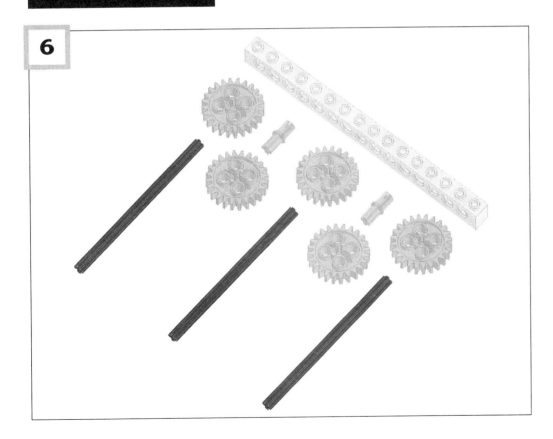

Gather a 1x16 beam, five 24-tooth gears, two axle pins, and three L10 cross axles.

7

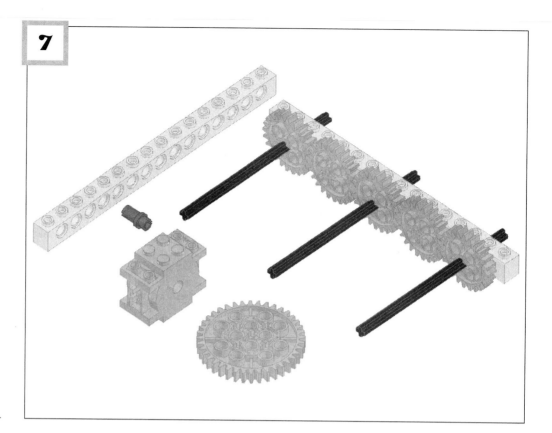

Attach the gears to the 1x16 beam as shown; mesh the teeth of the gears so that the angles of the cross axles are aligned (see the Note after Step 7 in Model 3 for what constitutes proper alignment).

8

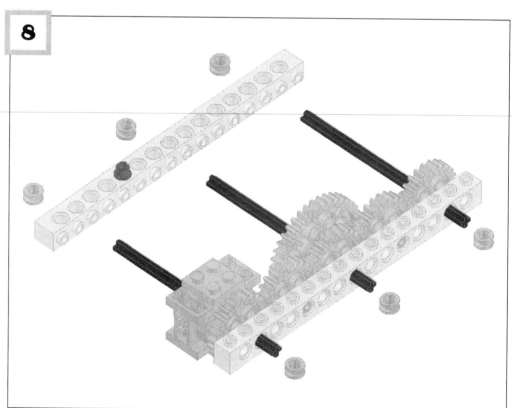

Slide a 40-tooth gear onto the middle cross axle. If you have a fiber optic light unit, slide it onto the end cross axle.

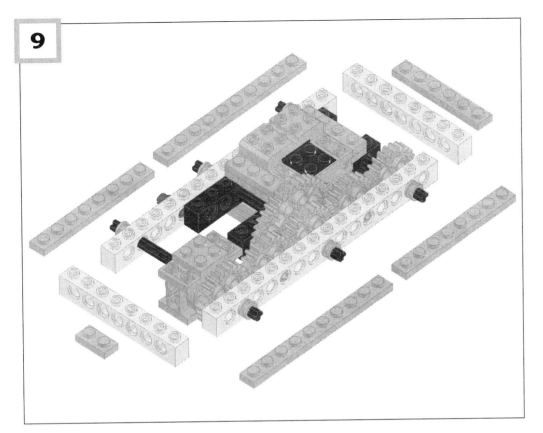

9

Pass the two cross axles on the right side of the figure through the motor assembly so that the 8-tooth gear on the motor meshes with the 40-tooth gear. Then slide a 1x16 beam onto the three cross axles and secure this beam by placing a half bushing on each cross axle.

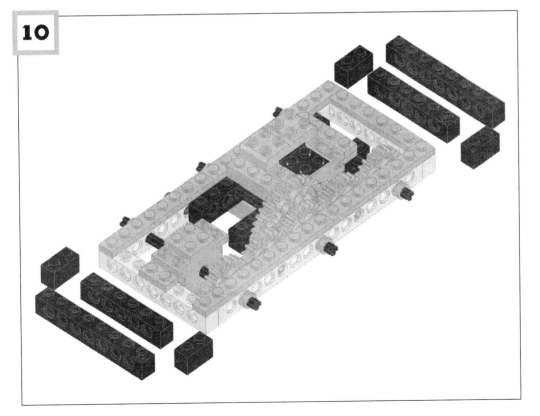

10

Reinforce the body by attaching 1x8, 1x10, and 1x12 plates on top of the beams that form the outer frame. Attach only a 1x2 plate on the front end next to the fiber optic light unit (shown at left in the figure) — we'll attach a light sensor here later.

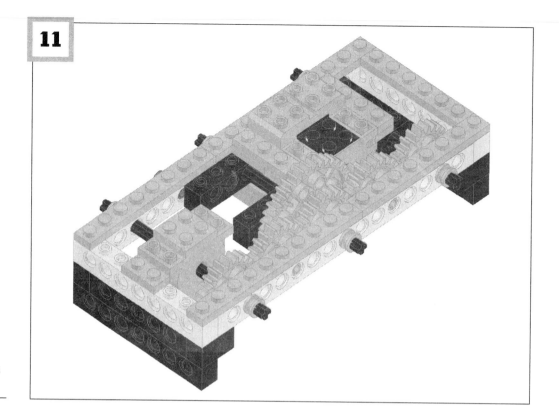

11

Attach 1x2 bricks and 1x6 beams to the bottom of the body and then attach 1x8 beams under them as shown.

BUILD THE STEERING SYSTEM

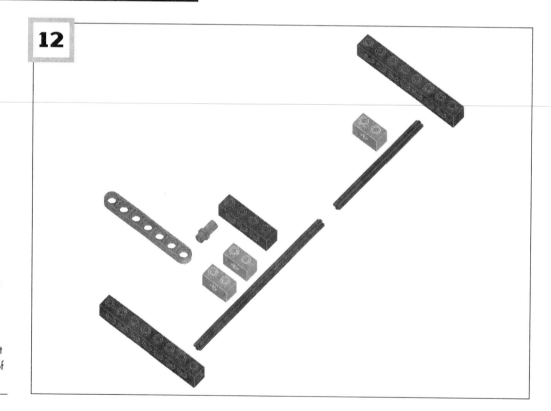

12

As discussed above, the Beetle will be able to turn left and right because the fulcrum of each leg can move up and down. Gather the pieces shown to build the steering system that will vary the position of each fulcrum.

13

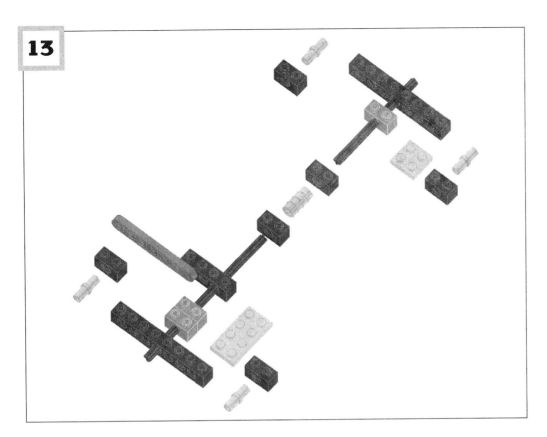

Slide two 1x2 beams onto the left side of an L12 cross axle, then slide one 1x2 beam onto the right side of an L8 cross axle. Pass each cross axle through a 1x8 beam. Next, slide a 1x4 beam onto the L12 cross axle and attach a 1x7 liftarm to the 1x4 beam with a 3/4 TECHNIC pin.

14

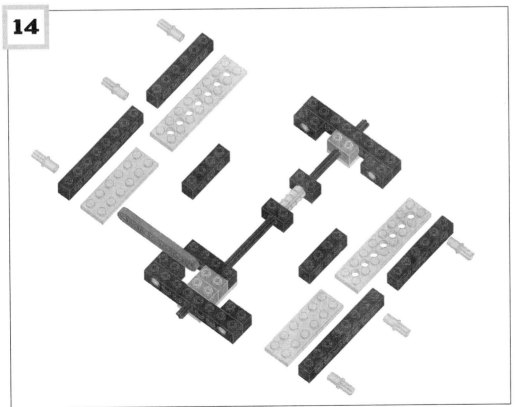

Attach a 2x4 plate under the L12 cross axle and a 2x2 plate under the L8 cross axle to secure them to the various beams through which they were passed. Next, slide a 1x2 beam onto each cross axle and connect the cross axles to each other with an axle joiner. Now use TECHNIC pins to attach a 1x2 beam to both ends of each 1x8 beam.

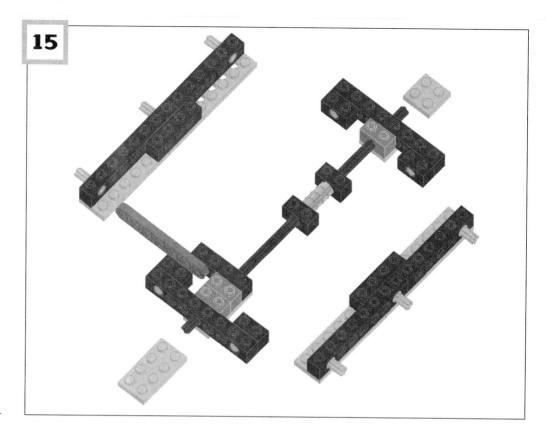

15

Combine 2x6 and 2x8 plates with 1x8 and 1x6 beams by arranging them in opposing orientation as shown; reinforce them by mounting 1x4 beams at their centers. Insert axle pins in the left, right, and center holes of these two structures.

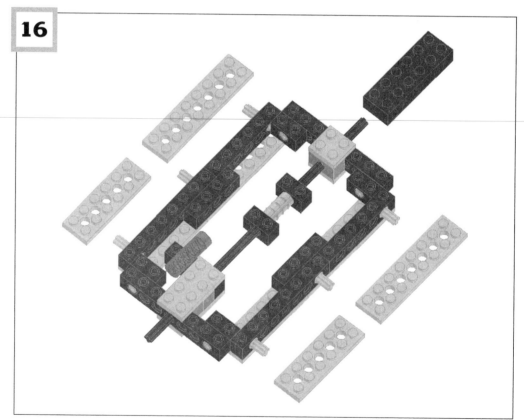

Add a 2x4 and 2x2 plate on top of the main structure for reinforcement, then attach the two structures built in Steps 12 through 15 to the 1x2 beams of the main structure, as shown.

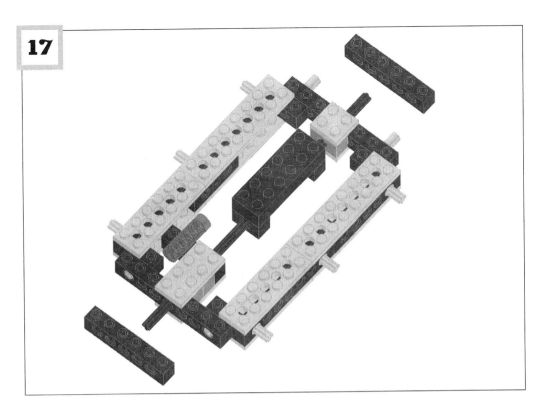

Use a 2x6 brick to connect the two 1x2 beams at the center of the axles. Attach 2x6 and 2x8 plates to the top of the long sides to complete the steering system.

ATTACH THE STEERING SYSTEM TO THE BODY

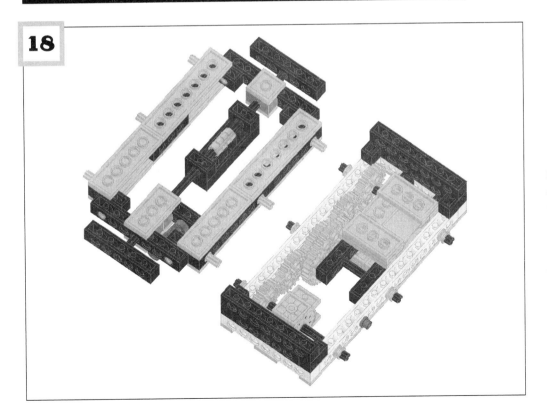

Now we'll attach the steering system to the bottom of the body—proceed carefully because this is a complicated assembly. It will be easier to proceed if you first turn both assemblies upside down. Attach 1x6 beams to the cross axles extending from both ends of the steering assembly.

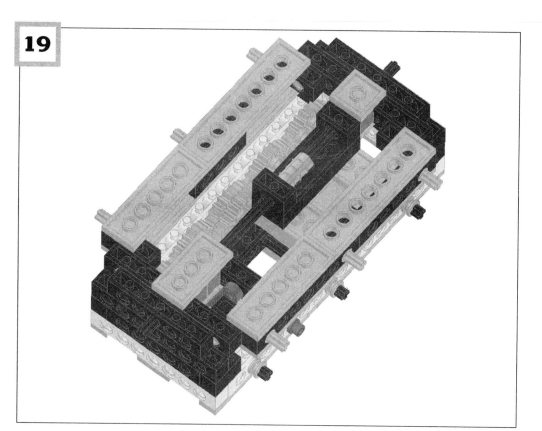

Attach the 1x6 beams added in Step 18 to the steering assembly to the 1x8 beams at either end of the body.

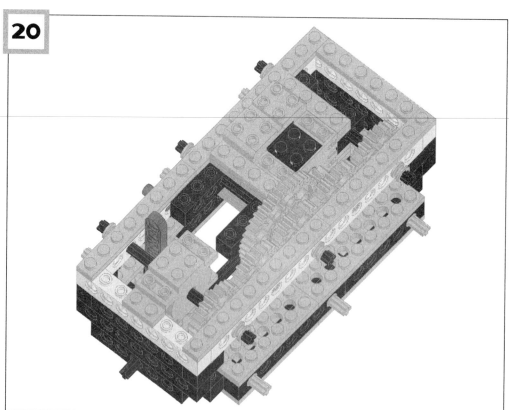

Turn the assembly over. The 1x7 liftarm should extend upward as in the figure.

21

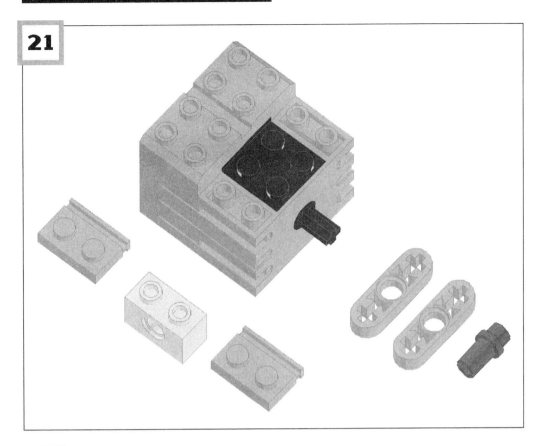

Gather the parts shown to attach a motor to the steering system.

22

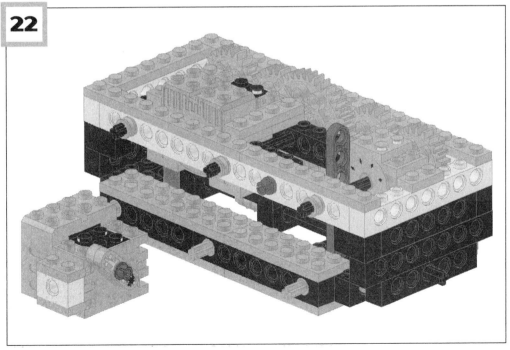

Attach 1x2 plates with rails and a 1x2 beam to one side of the motor. Slide two 1x3 liftarms onto the motor's cross axle and affix them by sliding a 3/4 TECHNIC pin through the center holes.

23

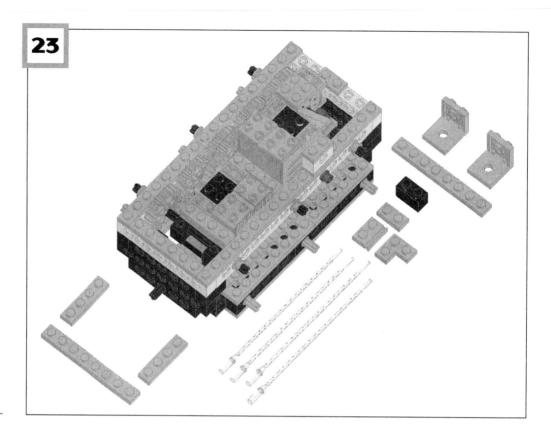

Mount the motor on top of the body and connect the 3/4 TECHNIC pin attached to the motor to the 1x7 liftarm extending up from the steering system.

24

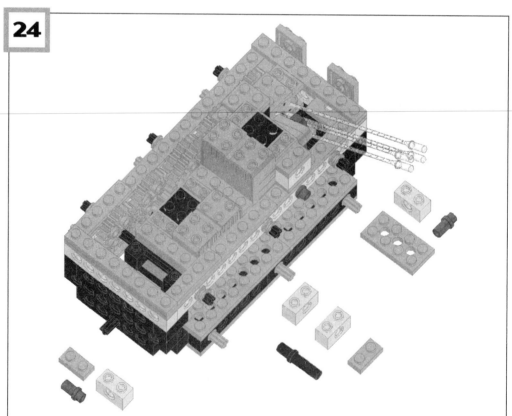

Attach a 1x2 plate with rail to the channel on the opposite side of the motor. Attach a 1x2 brick and plate to the top of the adjacent fiber optic light unit, and then secure these pieces with a 2x2 corner plate. Attach two 2x2–2x2 brackets to the front end of the assembly. Now attach the four fibers to the fiber optic light unit.

25

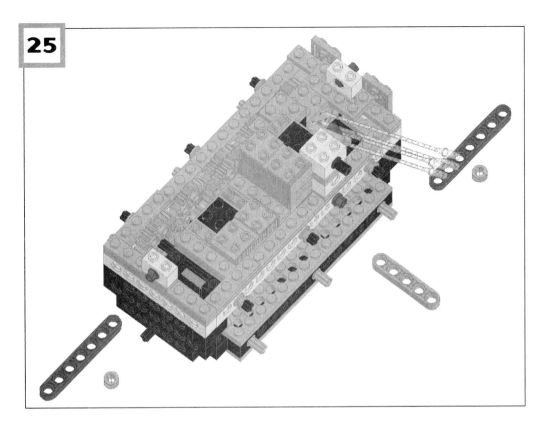

To strengthen the assembly, attach single 1x2 beams at the front and rear, and two near the motor. Insert a friction pin of length 3 into these two 1x2 beams.

26

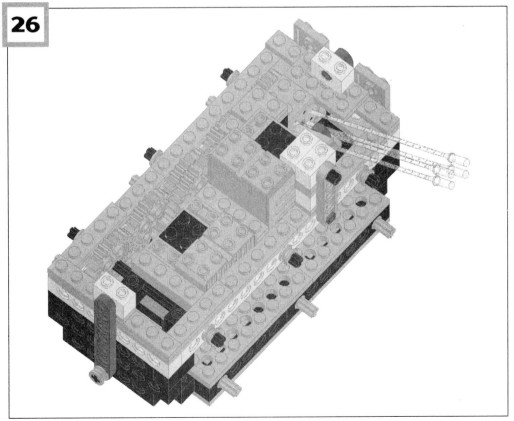

Attach 1x7 and 1x5 liftarms vertically to protruding cross axles below at all three sites with TECHNIC pins. Secure the 1x7 liftarms with half bushings. Securing the top and bottom portions of the assembly with these liftarms ensures that the robot stays together when it moves.

27

Gather the SLIZER parts shown to build the legs.

28

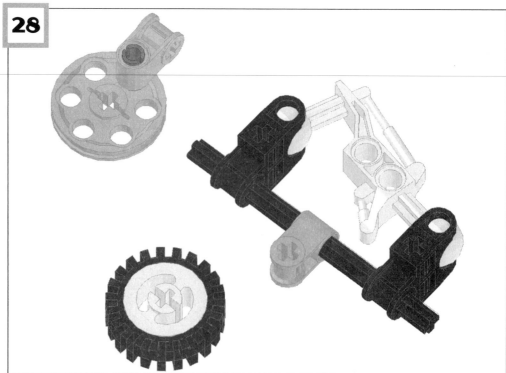

Slide a perpendicular axle joiner onto an L8 cross axle and secure both ends of the cross axle with SLIZER bearings. Attach a 3/4 TECHNIC pin to a pulley and then attach a perpendicular axle joiner to it.

29

Slide the perpendicular axle joiner attached to the pulley onto the top of the cross axle and attach a tire to the bottom of the cross axle to complete one left leg.

Now repeat Steps 27 through 29 to build two more left legs just like that in the figure and three right legs that are symmetrically reversed from the figure.

BUILD THE HEAD

30

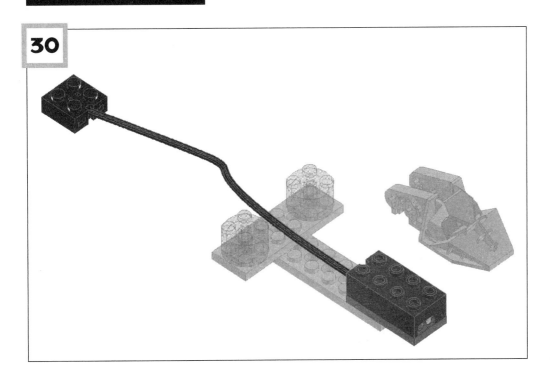

Build the head by combining a 2x6 and 2x10 plate and adding 2x2 rounded bricks and the light sensor, as shown.

31

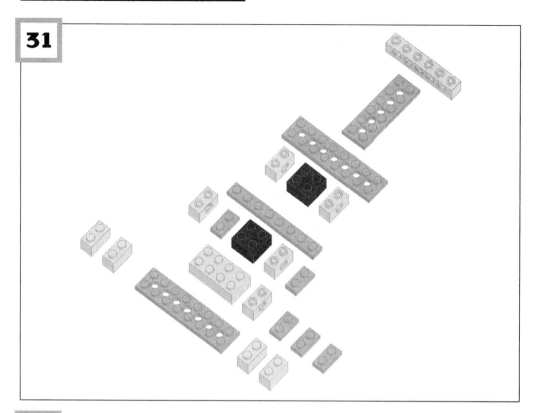

Gather the parts for building a base for the RCX, as shown.

32

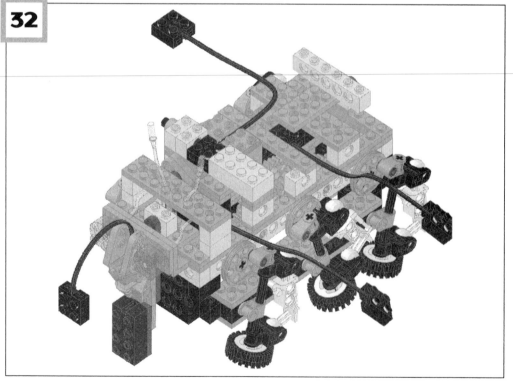

Secure these pieces on top of the body, then align and attach the six legs. Attach connecting leads to the motor and fiber optic unit and then attach the head with light sensor as shown. You can also add antennae if you wish.

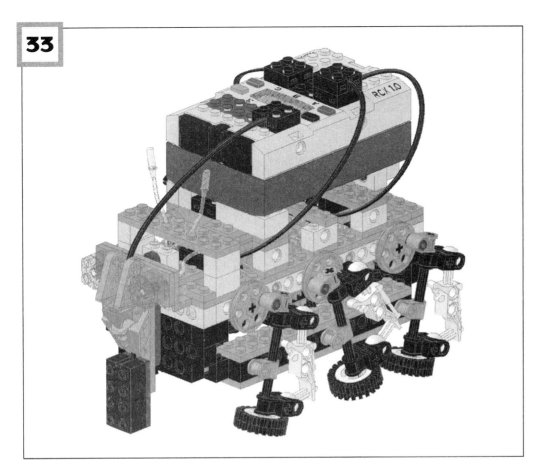

33

Mount the RCX and attach the connecting leads. Overlap the connecting leads from the motor set lower in the body (the one that moves the legs—set in Step 9) with those from the fiber optic unit and attach them to output A; attach the connecting lead from the motor resting on top of the body (the one that powers the steering—set in Step 23) to output C. Finally, attach the connecting lead from the light sensor to input 1.

PROGRAM

The NQC program below makes the Beetle follow a black line. When the Beetle senses a black area, it moves right. When it enters a white area, it moves left.

NOTE *To avoid applying an excessive load to the steering motor, set OUT_C to ON only once every 20 times through the loop.*

```
#define LEVEL 40
int direction;
int count;

task main()
{
        SetSensor(SENSOR_1, SENSOR_LIGHT);

        Wait(200);
        SetPower(OUT_A, OUT_FULL);
```

```
SetPower(OUT_C, OUT_FULL);
OnFwd(OUT_A);
direction=0;
while(true){
        if(SENSOR_1 < LEVEL && (direction != 1 || count > 20)){
                OnFwd(OUT_C);               //Go right
                Wait(5);
                Off(OUT_C);
                direction = 1;
                count=0;
        }

                if(SENSOR_1 > LEVEL && (direction != 1 || count > 20)){
                OnRev(OUT_C);               //Go left
                Wait(5);
                Off(OUT_C);
                direction = -1;
                count=0;

        }
        count++;
        }
    }
```

TIP

TWO BIRDS WITH ONE STONE

As you've seen, the more complicated the
robot, the more parts and motors are required.
Should you ever find that you're short of
motors to complete a project, you can use this
method to obtain two outputs from a single
motor. However, because this method isn't
magic, it's somewhat limited.

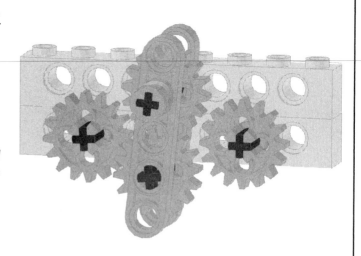

To do this, assemble a cog wheel as shown
below so that the lower middle gear, supported by liftarms, moves a little to the left and
right. When the motor rotates the middle cross
axle clockwise, the left gear rotates as shown;
when the middle cross axle is rotated counter-
clockwise, the right gear rotates.

This trick can be used, for example, to build a
single-motor robot that will follow a black line.

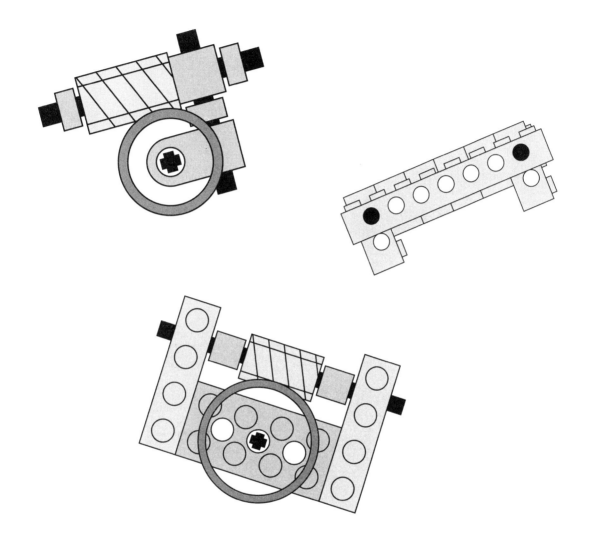

#924

#3005

#3648

#3010

#52009

METHODS OF MESHING GEARS

Various types and sizes of gears can be used in LEGO MINDSTORMS projects.
Following are ways to mesh these gears properly.

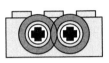

Two 8-tooth gears
Gear ratio: 1:1

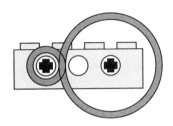

8-tooth and 24-tooth gear
Gear ratio: 1:3

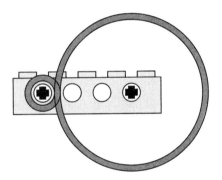

8-tooth and 40-tooth gear
Gear ratio: 1:5

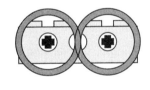

Two 16-tooth gears
Gear ratio: 1:1

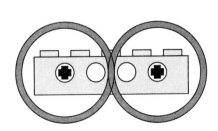

Two 24-tooth gears
Gear ratio: 1:1

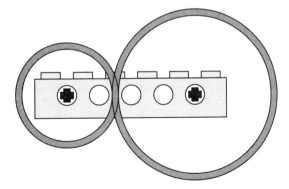

24-tooth and 40-tooth gear
Gear ratio: 3:5

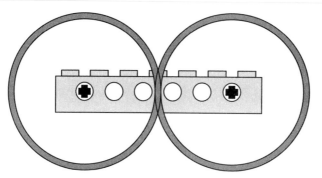

Two 40-tooth gears
Gear ratio: 1:1

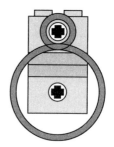

8-tooth and 24-tooth gear
Gear ratio: 1:3

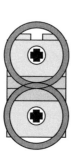

Two 16-tooth gears
Gear ratio: 1:1

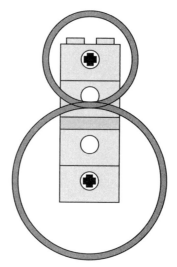

24-tooth and 40-tooth gear
Gear ratio: 3:5

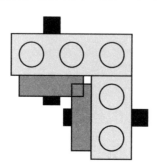

Two 12-tooth bevel gears
Gear ratio: 1:1

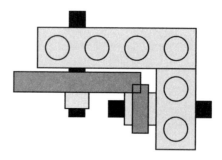

24-tooth crown gear and 8-tooth gear
Gear ratio: 3:1

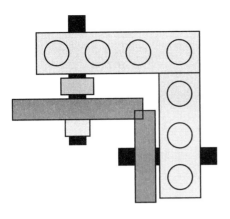

24-tooth crown gear and 16-tooth gear
Gear ratio: 3:2

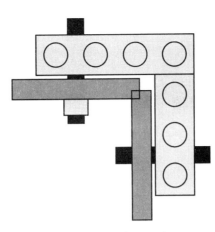

24-tooth crown gear and 24-tooth gear
Gear ratio: 1:1

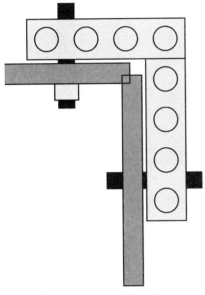

24-tooth crown gear and 40-tooth gear
Gear ratio: 3:5

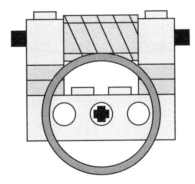

Worm gear and 24-tooth gear
Gear ratio: 1:24

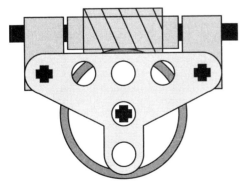

Worm gear and 24-tooth gear
Gear ratio: 1:24

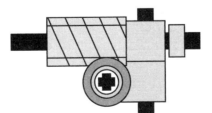

Worm gear and 8-tooth gear (using two perpendicular axle joiners)
Gear ratio: 1:8

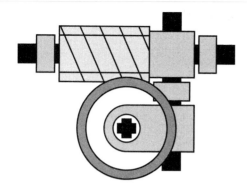

Worm gear and a 16-tooth gear (using two
perpendicular axle joiners)
Gear ratio: 1:16

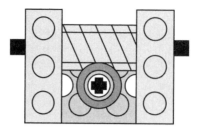

Worm gear and 8-tooth gear (the positions of
the gear and worm gear can also be reversed)
Gear ratio: 1:8

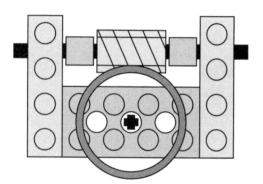

Worm gear and 24-tooth gear (the positions of the
gear and worm gear can also be reversed)
Gear ratio: 1:24

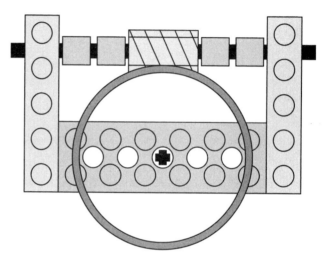

Worm gear and 40-tooth gear (the positions of the gear and worm
gear also can be reversed)
Gear ratio: 1:40

METHODS OF COMBINING BEAMS

If beams are combined perpendicularly, they produce a sturdy structure. However, to combine them perpendicularly, you must align their holes properly. The following shows standard ways to accomplish this.

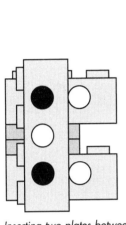

Inserting two plates between vertically stacked beams

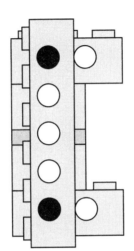

Inserting two beams and one plate between vertically stacked beams

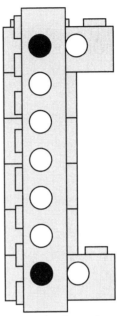

Inserting four beams between vertically stacked beams

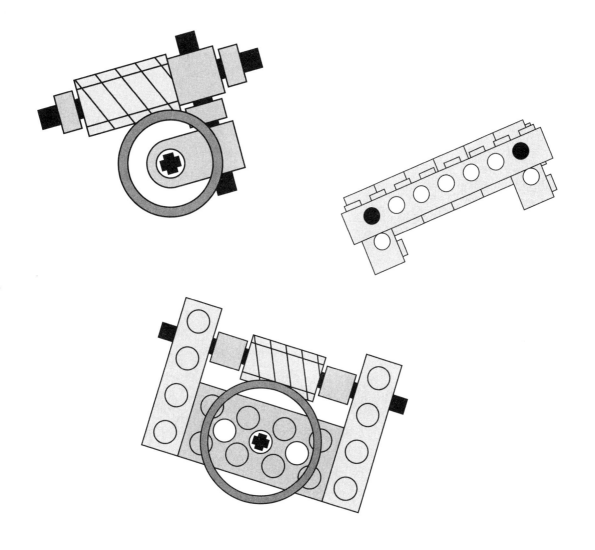

APPENDIX 2

NQC PROGRAMMING LANGUAGE

#924

#3005

#3648
#3010

#32009

The Not Quite C (NQC) programming language was developed by Dave Baum for the MINDSTORMS RCX. It provides more detailed control than RCX code does. Because NQC syntax is similar to that of the C programming language, anyone familiar with C programming should find NQC easy to use.

NQC is free software that can be downloaded from Dave Baum's Website (http://www.enteract.com/~dbaum/nqc/). The most recent release, NQC 2.2, can be used on a Macintosh, Windows, or Linux computer.

Following is a simple introduction to the syntax for NQC 2.2. Bold-faced items indicate actual syntax, while non-bold items are the variables to be filled in by the user. For information about the procedures for compiling NQC on Macintosh, Windows, and Linux computers, see the Web site mentioned above.

NOTE *The programs included in this book use NQC 2.2.*

TASKS

NQC 2.2	MEANING

task task_name( )
{
statement
}
	Defines a task. One task named **main** is always required in a program.

start task_name**;**	Starts the specified task.
stop task_name**;**	Stops the specified task.

INLINE FUNCTIONS

NQC 2.2	MEANING

void function_name(argument_list)
{
statement
{
Arguments:
   int variable
   const int variable
   int & variable
   const int & variable
	Defines a function that is expanded inline during compilation.
	
	Call by value (call using variable).
	Call by value (call using constant).
	Call by reference (call using variable).
	Call by reference (call using constant).

function_name(argument_list)**;** Arguments: Variables, constants	Calls the specified function.

SUBROUTINES

NQC 2.2	MEANING

sub subroutine_name( )
{
statement
}
	Defines a subroutine. A subroutine cannot be called from inside another subroutine because of RCX firmware limitations.

subroutine_name( );	Calls the specified subroutine.

VARIABLE DEFINITIONS

NQC 2.2	MEANING
int global_variable_name;	Defines a variable that can be used throughout the entire program.

{
   int local_variable_name;
}
	Defines a variable that can be used only in the relevant scope.

EXPRESSIONS

NQC 2.2	MEANING
variable **=** value;	Assigns value to variable.
variable **+=** value;	Adds value to variable.
variable **-=** value;	Subtracts value from variable.
variable ***=** value;	Multiplies variable by value.
variable **/=** value;	Divides variable by value.
variable **II=** value;	Sets variable to absolute value of value.
abs(value);	Absolute value.
variable **+-=** value;	Sets variable to sign of value **(0, +1, -1)**.
sign(value);	Sign of value **(0, +1, -1)**.
variable **&=** value;	Sets variable to bitwise **AND** of value.
variable **I=** value;	Sets variable to bitwise **OR** of value.
variable**++;** **++**variable;	Increments variable by 1.
variable**--;** **--**variable;	Decrements variable by 1.

The following operators can be used for constants or variables.

NQC 2.2	MEANING
+	Addition
-	Subtraction
*	Multiplication
/	Division
&	Bitwise **AND**
		Bitwise **OR**

The following operators can be used for constants.

NQC 2.2	MEANING
%	Modulo
<<	Shift left
>>	Shift right
^	Bitwise **XOR**
-	Minus (unary operator)
~	Complement of 1
&&	Logical **AND**
			Logical **OR**

RELATIONAL OPERATORS

NQC 2.2	MEANING
<	Less than
>	Greater than
<=	Less than or equal to
>=	Greater than or equal to
==	Equal to
!=	Not equal to
&&	Logical **AND**
			Logical **OR**
!	Negation

CONDITIONAL STATEMENTS

NQC 2.2	MEANING
if(cond_exp) statement	Executes statement if cond_exp is **true**.
if(cond_exp) statement_1 **else** statement_2	Executes statement_1 if cond_exp is **true** and executes statement_2 if cond_exp is **false**.

LOOPS

NQC 2.2	MEANING
while(cond_exp) statement	Repeatedly executes statement while cond_exp is **true**.
do statement **while**(cond_exp)	Repeatedly executes statement while cond_exp is **true**.
repeat(value) statement	Executes statement exactly value number of times.
until(cond_exp) statement	Repeatedly executes statement until cond_exp is **true**.

CONTROL STATEMENTS

NQC 2.2	MEANING
break;	Ends execution of innermost **while**, **do**, **repeat**, or **wait** and shifts control to next statement of the loop.
continue;	Shifts control to the end of the innermost **while**, **do**, **repeat**, or **wait** loop and proceeds with the next iteration of the loop.
return;	Ends a subroutine and shifts control to the calling statement.

FUNCTIONS

NQC 2.2	MEANING
SetSensor(in_port, setting)**;**	Configures a sensor.
in_port: **SENSOR_1, SENSOR_2,** **SENSOR_3**	
setting: **SENSOR_TOUCH**	Touch sensor (two values)
SENSOR_LIGHT	Light sensor (percent)
SENSOR_ROTATION	Angle sensor (angle)
SENSOR_PULSE	Touch sensor (pulse)
SENSOR_EDGE	Touch sensor (edge)
SENSOR_CELSIUS	Temperature sensor (Celsius)
SENSOR_FAHRENHEIT	Temperature sensor (Fahrenheit)
ClearSensor(in_port)**;**	Resets value of sensor.
in_port: **SENSOR_1, SENSOR_2,** **SENSOR_3**	
SetSensorType(in_port, type)**;**	Changes sensor type.
in_port: **SENSOR_1, SENSOR_2,** **SENSOR_3**	
type: **SENSOR_TYPE_TOUCH**	Touch sensor
SENSOR_TYPE_TEMPERATURE	Temperature sensor
SENSOR_TYPE_LIGHT	Light sensor
SENSOR_TYPE_ROTATION	Angle sensor
SetSensorMode(in_port, mode)**;**	Changes sensor mode.
in_port: **SENSOR_1, SENSOR_2,** **SENSOR_3**	
mode: **SENSOR_MODE_RAW**	Data from **0** to **1023**
SENSOR_MODE_BOOL	Boolean value (**true**, **false**)
SENSOR_MODE_EDGE	Edge
SENSOR_MODE_PULSE	Pulse
SENSOR_MODE_PERCENT	Percent
SENSOR_MODE_CELSIUS	Celsius
SENSOR_MODE_FAHRENHEIT	Fahrenheit
SENSOR_MODE_ROTATION	Angle (units: **360°/16**)
On(out_port)**;**	Rotates motor.
out_port: **OUT_A, OUT_B, OUT_C**	
OnFwd(out_port)**;**	Rotates motor in forward direction.
out_port: **OUT_A, OUT_B, OUT_C**	
OnRev(out_port)**;**	Rotates motor in reverse direction.
out_port: **OUT_A, OUT_B, OUT_C**	
OnFor(out_port, time)**;**	Rotates motor for fixed time.
out_port: **OUT_A, OUT_B, OUT_C**	Units: **10 ms**
time: variable	
Fwd(out_port)**;**	Sets motor rotation direction to forward
out_port: **OUT_A, OUT_B, OUT_C**	direction.

FUNCTIONS (CONT.)

NQC 2.2	MEANING
Rev(out_port); out_port: **OUT_A**, **OUT_B**, **OUT_C**	Sets motor rotation direction to reverse direction.
Toggle(out_port); out_port: **OUT_A**, **OUT_B**, **OUT_C**	Sets motor rotation direction to opposite direction.
Off(out_port); out_port: **OUT_A**, **OUT_B**, **OUT_C**	Stops motor.
Float(out_port); out_port: **OUT_A**, **OUT_B**, **OUT_C**	Causes motor to idle.

SetOutput(out_port, mode);
 out_port: **OUT_A**, **OUT_B**, **OUT_C**
 mode: **OUT_ON**
           OUT_OFF
           OUT_FLOAT
	Changes motor mode.
	
	 Rotate
	 Stop
	 Idle

SetDirection(out_port, dir);
 out_port: **OUT_A**, **OUT_B**, **OUT_C**
 dir: **OUT_FWD**
           OUT_REV
           OUT_TOGGLE
	Changes rotation direction of motor.
	
	 Forward
	 Reverse
	 Sets rotation direction to opposite direction.

SetPower(out_port, power);
 out_port: **OUT_A**, **OUT_B**, **OUT_C**
 power: **0** to **7**
           OUT_LOW
           OUT_HALF
           OUT_FULL
	Changes power of motor.
	
	 0
	 3 (**4** for **NQC 1.3**)
	 7

SYSTEM-DEFINED FUNCTIONS

NQC 2.2	MEANING
SensorValue(n); n: **0** to **2**	Returns sensor value.
SensorType(n); n: **0** to **2**	Returns sensor type.
SensorMode(n); n: **0** to **2**	Returns sensor mode.
SensorValueRaw(n); n: **0** to **2**	Returns sensor value as a number from **0** to **1023**.
SensorValueBool(n); n: **0** to **2**	Returns sensor value as a Boolean value (**true** or **false**).
Timer(n); n: **0** to **3**	Returns timer value.
Random(n);	Returns a random number from **0** to **n**.
Watch( );	Returns the elapsed time (in minutes) since the RCX was turned on.
Message( );	Returns the last **IR** message that was received.

DATALOG

NQC 2.2	MEANING
CreateDatalog(n); size: constant	Prepares **n** data areas.
AddToDatalog(data**);** data: variable, **Timer(x),** **SensorValue(x), Watch( )**	Adds value to datalog.

MISCELLANEOUS

NQC 2.2	MEANING
Wait(time); time: constant, variable, **Random(x)**	Halts for value X **10 ms**.
PlaySound(sound_type); sound_type: **SOUND_CLICK** **SOUND_DOUBLE_BEEP** **SOUND_DOWN** **SOUND_UP** **SOUND_LOW_BEEP** **SOUND_FAST_UP**	Plays sound. Key click Double beep From high tone to low tone From low tone to high tone Beep (error sound) From low tone to high tone quickly
PlayTone(freq, duration); freq: Tone frequency duration: Constant	Plays tone. Units: **10 ms**
SelectDisplay(display_mode); display_mode: **DISPLAY_WATCH** **DISPLAY_SENSOR_1** **DISPLAY_SENSOR_2** **DISPLAY_SENSOR_3** **DISPLAY_OUT_A** **DISPLAY_OUT_B** **DISPLAY_OUT_C**	Sets display contents of RCX LCD. System clock Input port **1** Input port **2** Input port **3** Output port **A** Output port **B** Output port **C**
SendMessage(message); message: constant or variable	Sends message to another RCX.
ClearMessage( );	Clears received messages.
ClearTimer(n); n: **0** to **3**	Resets timer.
StopAllTasks( );	Stops all tasks.
SetWatch(hours, minutes);	Sets RCX time as hours:minutes.
SetTxPower(IR_mode); IR_mode: **TX_POWER_LO,** **TX_POWER_HI**	Sets **IR** mode.

COMMENTS

NQC 2.2	MEANING
/* comment */	Treats everything between /* and */ as a comment.
// comment	Treats everything from // to the end of the line as a comment.

MACROS

NQC 2.2	MEANING
#include "file_name"	Expands contents of file into program.
#define identifier string	Replaces identifier within program with string. If string spans multiple lines, \ is appended at the end of each line.
#undef identifier	Disables definition by **#define**.
#if const_exp	If const_exp is false, causes everything until **#else, #elif,** or **#endif** to be ignored.
#ifdef identifier	If identifier is undefined, causes everything until **#else, #elif,** or **#endif** to be ignored.
#ifndef identifier	If identifier is defined, causes everything until **#else, #elif,** or **#endif** to be ignored.
#else	When corresponding **#if**, **#ifdef**, or **#ifndef** is true, causes everything until **#endif** to be ignored.
#elif const_exp	Use to assign an additional **#if** within an **#else** macro.
#endif	Indicates end of **#if** macro.
#pragma noinit	Causes RCX not to be initialized.
#pragma init func_name	Performs user-defined initialization.
defined(identifier)**;**	Indicates whether identifier is defined by **#defined** statement. Can be used in macro definition expression.

JIN SATO'S LEGO® MINDSTORMS™: THE MASTER'S TECHNIQUE

by JIN SATO

In his landmark book, Jin Sato introduces the basic principles of robotic engineering, including how to plan and build robots with tires, legs, and grasping hands. Readers even learn how to build Sato's famous robotic dog MIBO as well as four other robots. Includes a parts list for the robots and coverage of the programming environment, assembly drawings, and maintenance of robots.

AUGUST 2001, 200 pp., four color insert, $29.95 ($44.95 CDN)
ISBN 1-886411-56-5

ASTRONOMER'S COMPUTER COMPANION

by JEFF FOUST *and* RON LAFON

A professional astronomer and an avid amateur astronomer have teamed up to introduce basic concepts of astronomy and explain how to find and use software to aid in observation. The contains software, astronomical images and animations, and a searchable electronic version of the book.

1999, 384 pp. W/CD-ROM, $39.95 ($61.00 CDN)
ISBN 1-886411-22-0

THE LITTLE RED BOOK OF ADOBE LIVEMOTION

by DEREK PELL

By purging technical jargon and attacking LiveMotion from a working artist's perspective, former National Lampoon writer Derek Pell offers geek-free simplicity and a refreshingly radical approach to design. This engaging lampoon reveals how to use LiveMotion to produce Flash animations; add and mix audio files; build preloaders; and much more.

JUNE 2001, 200 pp., $19.95 ($29.95 CDN)
ISBN 1-886411-53-0

STEAL THIS COMPUTER BOOK 2

by WALLACE WANG

This offbeat, non-technical book is for any computer user interested in Internet security issues like viruses, cracking, and password theft in the same informative, irreverent, and entertaining style that made the first edition a huge success. The CD-ROM contains hundreds of anti-hacker and security tools for Windows, Macintosh, and Linux.

2000, 400 pp. W/CD-ROM, $24.95 ($38.95 CDN)
ISBN 1-886411-42-5

THE BLENDER BOOK

by CARSTEN WARTMANN

The Blender Book's step-by-step tutorials demystify Blender's complex interface and show how to use all aspects of Blender to enhance and animate Web sites, graphics designs, and video productions; use materials and textures; work with skeleton animation and kinematics; and integrate 3D objects into videos. The CD-ROM contains Blender for all platforms, as well as textures and all tutorials, scenes, and animations from the book.

2001, 350 pp. W/CD-ROM, $39.95 ($59.95 CDN)
ISBN 1-886411-44-1

Phone:
1 (800) 420-7240 OR
(415) 863-9900
MONDAY THROUGH FRIDAY,
9 A.M. TO 5 P.M. (PST)

Fax:
(415) 863-9950
24 HOURS A DAY,
7 DAYS A WEEK

E-mail:
SALES@NOSTARCH.COM

Web:
HTTP://WWW.NOSTARCH.COM

Mail:
NO STARCH PRESS
555 DE HARO STREET, SUITE 250
SAN FRANCISCO, CA 94107
USA

Distributed to the book trade by Publishers Group West

UPDATES

This book was carefully reviewed for technical accuracy, but it's inevitable that some things will change after the book goes to press. Visit **http://www.nostarch.com/lmib_updates.htm** for updates, errata, and other information.

FIND ALL IMAGES FROM
THIS BOOK ONLINE!

If you'd like to examine any of the images in this book more closely, you'll find them online at **http://www.nostarch.com/?robotics**. You'll also find links to LEGO modeling software, alternative programming languages (like NQC), movies of MINDSTORMS robots, and updates on forthcoming publications about MINDSTORMS products from No Starch Press.

See you there!